SpringerBriefs in Applied Sciences and Technology

Thermal Engineering and Applied Science

Series Editor

Francis A. Kulacki

For further volumes:
http://www.springer.com/series/10305

Aniruddha Bagchi · Francis A. Kulacki

Natural Convection in Superposed Fluid-Porous Layers

 Springer

Aniruddha Bagchi
Veeco Process Equipment Inc.
Somerset, NJ
USA

Francis A. Kulacki
Department of Mechanical Engineering
University of Minnesota
Minneapolis, MN
USA

ISSN 2193-2530   ISSN 2193-2549 (electronic)
ISBN 978-1-4614-6575-1 ISBN 978-1-4614-6576-8 (eBook)
DOI 10.1007/978-1-4614-6576-8
Springer New York Heidelberg Dordrecht London

Library of Congress Control Number: 2013940311

Printed on acid-free paper

Springer is part of Springer Science+Business Media (www.springer.com)

Preface

Natural convection heat transfer in domains containing superposed fluid and porous layers is a fundamental transport mechanism encountered in a wide range of engineering, geophysical, and scientific applications. In this framework, the porous sub-layer lies below a saturating liquid layer. Important engineering applications include packed bed solar energy storage, directional solidification of binary alloys, fibrous and granular insulation systems, water reservoirs, and post-accident cooling of nuclear reactors. Thermal circulation in lakes and shallow coastal areas and contaminant transport in groundwater represent some of the geophysical applications. Although buoyant convection in this system was first studied about 40 years ago, there has lately been renewed interest in this problem owing to its importance in environmental and energy management problems in current scientific and geo-political contexts.

A major area of recent investigation has been the reconciliation of the several modeling approaches currently in use. Of particular interest are the boundary conditions at the porous fluid-layer interface and the prediction of overall heat transfer coefficients when convective currents are driven by a heat source of finite size on the lower boundary. While different modeling approaches give near-identical solutions for vertically layered systems, they disagree significantly for horizontal systems. Therefore, a significant amount of research has been aimed at identifying these issues and developing an accurate and consistent mathematical formulation. The dearth of measurements provides no validating experiments.

We have undertaken the present topic as it has neither been extensively explored nor resolved. A combination of numerical analysis and experiments has led us to this first comprehensive summary report and the first report of measurements of convective heat transfer coefficients well above the onset point of convection. This work also represents a step toward validation of numerical predictions in the high Rayleigh number range. Such validation of theoretical studies is virtually absent in the literature owing to the difficulty encountered in designing experiments which allow heat transfer measurement in the Rayleigh number range accessible to simulation. Although numerical solutions of this problem were first published over 25 years ago, no accepted set of results is yet available even for the case of a uniformly heated base. This is especially true for the high Rayleigh number regime where the lack of experimental validation has made it impossible to verify the accuracy of any of the modeling approaches.

Similarly, experimental studies have focused on studying very small sections of the entire convective heat transfer regime, partly because of the complexity of the dual layer problem and the large number of independent controlling parameters. Part of the reason that the overall nature of the problem is not yet well understood is that previous studies have focused on studying very specific aspects of it. Taken in this context, the combined numerical-experimental approach here represents an important step toward fundamentally understanding the problem over a large portion of the entire heat transfer regime. Also, we have considered in-depth an extension of the superposed layer problem by investigating the effects of localized heat sources which are more relevant from an engineering perspective.

Somerset, NJ, USA   Aniruddha Bagchi
Minneapolis, MN, USA   Francis A. Kulacki

Contents

Symbols

a	Wave number
A	Aspect ratio, L/H
c_p	Heat capacity, J/kg K
d	Particle diameter, m
Da	Darcy number, K/H^2
Da_p	Darcy number, K/d^2
F	Forchheimer coefficient, $1.75/(175^{1/2}\phi^{3/2})$
g	Gravitational acceleration, (0,-g), m/s^2
H	Height of the problem domain, m
h	Heat transfer coefficient, W/m^2
i, j	Unit vectors in x- and y-directions
k	Thermal conductivity, W/m K
K	Permeability, m^2
L	Length of the problem domain, m
N	Nusselt number, hH/k_f
Nu_m	Nusselt number based on the stagnant conductivity, hH/k_m
q	Heat supplied at the heater, W
q_{loss}	Heat loss, W
q''	Heat flux, W/m^2
P	Pressure, N/m^2
Pr	Prandtl number, ν_f/α_f
Ra	Rayleigh number, $g\beta H^3(T_H - T_C)/\nu_f\alpha_f$
Ra_m	Rayleigh number for the porous layer, $g\beta KH(T_H - T_C)/\nu_f\alpha_m$
T	Temperature, K
t	Time, s
u	Velocity, (u,v), m/s
x, y	Cartesian co-ordinates, m

Greek Letters

ϕ Porosity
$\hat{\alpha}$ Beavers–Joseph constant
α Thermal diffusivity, m^2/s
β Coefficient of thermal expansion of the fluid, $1/K$
$\hat{\beta}$ Stress-jump coefficient
γ Non-dimensional particle diameter, $d/H^{\text{'}}$
δ Heater-to-base length ratio, L_H/L
ε Thermal diffusivity ratio, α_f/α_{so}
ζ Vertical through flow strength, $v_t H_m/\alpha_m$
η Ratio of porous layer height to total height, H_m/H
η_1 Ratio of fluid layer height to porous layer height, H_f/H_m
Θ Vector of quantities in discretized equations
κ Effective conductivity ratio, k_m/k_f
λ Thermal conductivity ratio, k_{so}/k_f
ν Kinematic viscosity, m^2/s
ξ Horizontal-to-vertical permeability ratio in an anisotropic porous layer
Ξ Viscosity variation parameter, $\log(\nu_{max}/\nu_{min})$
ρ Density, kg/m^3
σ Effective heat capacity ratio, $(\rho c_p)_m/(\rho c_p)_f$
χ Horizontal to vertical thermal diffusivity ratio in an anisotropic porous layer
ψ Stream function, m^2/s
ω Vorticity, s^{-1}

Subscripts

avg   Average
0   Reference
C   Cool wall
c   Critical
eff   Effective
f   Fluid
H   Hot wall
m   Porous layer
max   Maximum
min   Minimum
so   Solid

Superscripts

*   Dimensionless

Chapter 1
Introduction

Keywords Natural convection • Rayleigh-Bénard convection • Porous media • Fluid-superposed porous layers

A porous medium can be loosely defined as a solid with an interconnected void space. Figure 1.1a shows a schematic of an idealized porous medium comprising uniform spherical particles saturated with a single fluid. In general, porous media, especially those that occur naturally, have an irregular geometry as shown in Fig. 1.1b. Porous media are ubiquitous in nature and can also be found in several engineering applications. Examples of naturally occurring porous media include beach sand, sandstone, limestone, rye bread, wood, and the human lung, while engineering applications include packed bed reactors, geothermal energy extraction, energy storage devices, and thermal insulation systems.

Owing to the wide range of situations in which porous media are encountered, the study of heat transfer in these systems has received a great deal of attention in the scientific and engineering communities. A problem that has attracted significant attention is that of natural convection in horizontal porous layers uniformly heated from below. Beginning with the pioneering studies of Horton and Rogers (1945) and Lapwood (1948), this problem, which is the porous medium equivalent of the Rayleigh-Bénard problem, has been studied extensively. Several variations of this problem have also been studied in great detail. These include convection in multi-layered porous media, convection with localized heat sources, and convection in anisotropic porous media. These studies have been motivated by practical applications, such as the disposal of high level nuclear wastes in deep geological repositories, cooling of electronic devices, design of energy efficient buildings, and geothermal energy extraction. The problem also gains importance from a fundamental scientific perspective as this is an example of a system where a well-defined flow structure develops from an initial random disturbance and therefore allows for a fundamental investigation of stability modes. Additionally, the mathematical formulation of the problem is one of the simplest nonlinear elliptic systems. As a

A. Bagchi and F. A. Kulacki, *Natural Convection in Superposed Fluid-Porous Layers*, SpringerBriefs in Thermal Engineering and Applied Sciences, DOI: 10.1007/978-1-4614-6576-8_1, © The Author(s) 2014

Fig. 1.1 a An idealized
saturated porous medium;
b a naturally occurring
porous medium

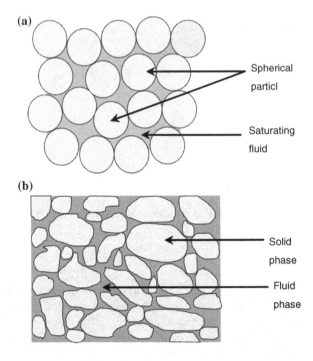

result, it has been an ideal test problem for numerous numerical studies. Reviews by Combarnous and Bories (1975) and Cheng (1978) give detailed accounts of the early advances in the field, and books by Nield and Bejan (2006) and Kaviany (1991) contain comprehensive overviews of the then current state of research.

An important variation of this problem is that of natural convection in a horizontal composite system comprising a fluid layer overlying a porous layer saturated with the same fluid. This problem is of great importance from a geophysical perspective. Mixing in ice covered lakes (Matthews 1998), flow in geysers and hot springs, flow of oil in underground reservoirs (Allen 1984; Ewing 1996), patterned ground formation under water, (Carr and Straughan 2003) and contaminant transport in sub-soil water reservoirs (Curran and Allen 1990, Allen and Khosravani 1992) are all natural examples. The problem also arises in several engineering and industrial applications such as fibrous and granular thermal insulation systems, water reservoirs, grain storage installations, hydrothermal synthesis in the growth of crystalline materials (Chen et al. 1999), solidification of alloys (Worster 1992), and post-accident cooling of nuclear reactors (Rhee et al. 1978). In spite of this universality, the fluid-superposed porous layer convection problem has not been investigated in great depth.

Most existing studies of the topic deal with the onset of convection in the composite system heated uniformly from below (Sun 1973; Nield 1977; Somerton and Catton 1982; Chen and Chen 1988, 1989, Chen and Hsu 1991; Chen et al. 1991; Chen and Lu 1992; Zhao and Chen 2001; Hirata and co-workers 2006–2009). These studies have identified several important aspects of this problem, the most important being that convection in such a composite system has a bimodal character and that the wavelength of the convective mode at the onset point depends on the fluid-to-porous layer height ratio, η_1. Even then, existing studies do not agree on the exact critical point primarily because there is no agreement on the most accurate approach to modeling fluid motion in this system. In particular, there has been extensive debate on the appropriate form of the boundary conditions at the porous-fluid layer interface. Three different modeling approaches have been proposed, and all of them predict somewhat different critical onset points. Recently there has been much work on reconciling the different models and identifying reasons for the observed discrepancy of the results (Zhao and Chen 2001; Hirata and co-workers 2006–2009). The lack of conclusive experimental studies has led to the problem being, as yet, unresolved.

Similarly, the overall heat transfer characteristics of the system at a high Rayleigh numbers are not well understood. Only a handful of studies have explored this problem from a theoretical perspective (Poulikakos and co-workers 1986; Chen and Chen 1992; Kim and Choi 1996), and they report widely different results. This can, in part, be attributed directly to the different modeling approaches adopted in these studies. A comprehensive set of experimental studies by Prasad and co-workers (Prasad and Tian 1990; Prasad et al. 1991; Prasad 1993) are available, but they have not been confirmed by other investigators. Moreover, no direct comparison of experimental and theoretical results at a high Rayleigh numbers has ever been published and this further compounds the difficulty in validating numerical studies.

A major area of recent investigations has been the reconciling of several modeling approaches that are available for theoretically studying the problem. In addition, there has been some progress toward validating numerical predictions with experimental results and developing the Nusselt-versus-Rayleigh number correlations over a wide range of parameters. Therefore the literature on this topic is evolving continuously as newer studies keep exploring various aspects of this problem in greater detail.

The aim of this monograph is to present an up-to-date report on the current state of research on the problem of natural convection in horizontal composite fluid-porous domains. As mentioned earlier, this problem was first studied nearly 40 years ago and since then, there has been an ever expanding body of literature. Therefore a need was felt to summarize the various aspects of this problem in a single volume that provides a comprehensive discourse on it. Similar attempts at summarizing this problem in the past have focused on giving a detailed literature review of past studies (Prasad 1991; Gobin and Goyeau 2008). In this monograph, a more comprehensive overview is presented by discussing in detail the current literature, underlying theoretical principles, controlling parameters, numerical solution techniques, and recent experimental investigations. A horizontal composite domain heated locally from below is selected as a test case for illustrating the various aspects of this problem. Considering the wide range of applications where this problem is encountered, we have presented the subject in a way that is accessible to audiences in a variety of scientific and engineering backgrounds.

Chapter 2
Literature Review

Keywords Convective instability • Critical point • High Rayleigh number convection • Bi-modal convection • Short wavelength convection mode • Long wavelength convection mode

In this chapter, the current status of research is presented. The progression of solutions includes the problem of predicting the onset of convection for an initially motionless state to the calculation of velocity and temperature fields for two-dimensional steady convection.

2.1 Current Status of Research

The prototypical problem for natural convection in horizontal fluid-superposed porous layers can be represented as shown in Fig. 2.1. A saturated porous layer with an overlying fluid layer is held between two impermeable surfaces. The lower surface is held at a higher temperature than the upper surface and side walls are adiabatic. The most commonly studied case is where the high temperature boundary is maintained at a constant temperature. A few studies have also examined the case wherein the lower boundary is either heated by constant heat flux or when the underlying fluid layer is volumetrically heated.

A primary goal of the research in this field is to calculate the critical Rayleigh number and to understand the relation between the Nusselt and the Rayleigh numbers beyond the critical point. The Rayleigh number is a dimensionless parameter that defines the strength of the buoyancy force that drives fluid motion during natural convection. When the value of the Rayleigh number is below a critical value, there is no fluid motion and conduction is the dominant mode of heat transfer. Beyond the critical Rayleigh number, convection becomes the dominant mode of heat transfer due to buoyancy. On the other hand, the Nusselt number is

A. Bagchi and F. A. Kulacki, *Natural Convection in Superposed Fluid-Porous Layers*,
SpringerBriefs in Thermal Engineering and Applied Sciences,
DOI: 10.1007/978-1-4614-6576-8_2, © The Author(s) 2014

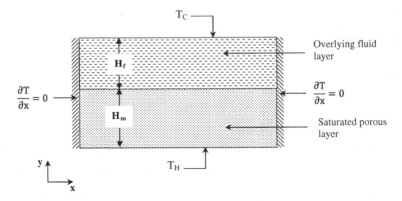

Fig. 2.1 Problem domain for the composite layer heated uniformly from below. $T_H > T_C$

a dimensionless parameter representing the ratio of convective to conductive heat transfer. The Nusselt number is therefore a measure of the increase in the rate of heat transfer due to fluid motion.

Sun (1973) was probably the first to examine the onset of convection in this composite system and conducted a comprehensive analytical and experimental study of convective stability. Using the perturbation method, he predicts the critical Rayleigh number for a variety of thermal and hydrodynamic conditions at the bounding surfaces. He considers the influence of various parameters such as the layer depth ratio, porosity, permeability, density, and diffusivity and finds that the critical Rayleigh number for various bed height ratios compares well with his experimental data for $0.82 < \eta < 1$. He also finds that for a fixed Rayleigh number, the Nusselt number decreases with an increase in the height of the porous bed.

This problem was later studied analytically by Nield (1977) who uses linear stability theory to predict the onset of convection in a two-layer system that extends infinitely in the horizontal direction. He applies the Beavers-Joseph (1967) condition at the fluid-porous layer interface and includes the possibility of the Marangoni effect at a deformable upper surface. Nield, however, does not solve the resulting tenth-order eigenvalue problem because of the tedious nature of the required solution. Instead he considers the simplest case of constant heat flux at the lower boundary and obtains the stability criterion for limiting values of various parameters, such as the fluid viscosity, height ratio, and conductivity ratio.

Rhee, Dhir and Catton (1978) measure heat transfer rates in a composite system consisting of a porous layer made up of heat generating particles cooled by a layer of fluid from above. Their experiments are conducted in a cylindrical test section with 6.35 mm diameter steel balls as the porous layer and distilled water as the saturating fluid. The sides and the bottom of the test section are kept insulated while the top surface is maintained at a constant temperature. The steel balls are heated by an induction coil that is placed around the lower section of the test section. Results show that the critical Rayleigh number for the porous layer, Ra_{mc}, decreases rapidly as $\eta_1 \to 1$, and above this value of η_1, $Ra_{mc} \to 12$

asymptotically. In the absence of the overlying liquid layer, however, the critical Rayleigh number is 46, indicating that the presence of the liquid layer facilitates the onset of convection. The presence of the liquid layer is also seen to increase the rate of heat transfer through the bed. For a liquid layer about one-fifth the height of the porous layer, there is a three-fold increase in the heat transfer coefficient. They note that this is likely due to the inflow and outflow of fluid from the porous bed. They also observe that for $\eta_1 > 1$, heat transfer data can be correlated with $Nu_m = 0.19\, Ra_m^{0.69}$.

The problem of a fluid layer overlying a volumetrically heated porous layer is investigated analytically by Somerton and Catton (1982), who conduct a linear stability analysis to predict the onset of convection. Brinkman's extension of Darcy's law is used to model fluid flow in the porous layer, which allows them to avoid using the Beavers-Joseph boundary condition at the interface and instead use more generalized boundary conditions describing the continuity of velocity and shear stress. They find: (i) the critical wave number is a function of the fluid-to-bed depth ratio, η_1, the conductivity ratio, λ, and the Darcy number, Da; (ii) the sole parameter controlling the onset of convection is a single Rayleigh number for the entire system; (iii) a larger λ tends to a more stable fluid layer while a large Da tends to produce a less stable fluid layer; and (iv) the presence of the overlying fluid layer is destabilizing and can drive fluid motion in the porous bed. However, they do not calculate the heat transfer characteristics of the system.

Poulikakos, Bejan, Selimos and Blake (1986) solve the problem of flow instability in a horizontal composite layer beyond the critical Rayleigh number. They numerically solve the full set of conservation equations for two-dimensional transient flow and use the Beavers-Joseph boundary condition at the interface for fixed values of the Prandtl number (Pr = 7), thermal diffusivity ratio ($\varepsilon = 1$) and porous-to-total layer height ratio ($\eta = 0.5$). The parameters varied are the aspect ratio, A, (0.4–2), fluid Rayleigh number, Ra, (10^2–10^6) and Darcy number, Da (10^{-7}–10^{-4}). They obtain streamlines and isotherms for various values of these parameters and find that with increasing Rayleigh number, the velocities of the observed two-dimensional cellular rolls increase but their numbers remain fixed within the horizontal extent of the computational domain. The authors do not observe any significant change in the Nusselt number with an increase in the Darcy number but find that it decreases with increasing aspect ratio. They also find that the critical fluid Rayleigh number is in the range of 500–600. Beyond this range, the Nusselt number is given by $Nu \approx 0.129 Ra^{0.33}$.

Poulikakos (1986) extends the above study by using a general flow model that incorporates the Brinkman and Forchheimer extensions to the Darcy model and a single set of conservation equations for both the fluid and porous layers. This duality is accomplished via a parameter which assumes the value zero in the fluid region, unity in the porous region, and avoids the explicit specification of boundary conditions at the porous-fluid layer interface. Numerical solutions are obtained for two-dimensional flow by using a finite-volume technique for various values of parameters such as the aspect ratio, A, (1–5), height ratio, η, (0.1–0.8), fluid Rayleigh number, Ra (10^3–10^5), the Darcy number, Da (10^{-5}–10^{-3}), and Forchheimer coefficient, F (0–1.0). Poulikakos finds that the velocity of convective motion increases with an increase in the Rayleigh number when all other parameters are held constant and that the number of cells doubles when the Rayleigh number attains the value of 2×10^5. With an increase in the height of the overlying fluid, a significant increase in the overall Nusselt number occurs. Beyond a critical height ratio of $\eta \approx 0.75$, the flow field ceases to have any effect on heat transfer and heat removal is primarily by conduction. An increase in Darcy number also increases the Nusselt number.

In a series of studies, Chen and co-workers investigate the convective instabilities and heat transfer characteristics of fluid-superposed porous layers, both analytically and experimentally. In connection to the problem of directional solidification of concentrated alloys, Chen and Chen (1988) first consider the problem of salt-finger convection and use linear stability analysis to determine the onset of finger convection. Darcy's law is used to model fluid flow in the porous layer, and the Beavers-Joseph condition is applied at the fluid-porous layer interface. The additional effect of a salinity gradient is included in the conservation equations to model salt-fingering. They find that the parameters which affect the onset of flow are the fluid-to-porous layer height ratio, the Darcy number, the thermal diffusivity ratio, and the Beavers-Joseph constant. For fixed values of Da, ε, and $\hat{\alpha}$, they find that the marginal stability curve is bimodal. For $\eta_1 < 0.12$, the long wave branch is most unstable and convection is dominated by the porous layer. For $\eta_1 > 0.12$, the short wave branch is the most unstable and convection is dominated by the fluid layer. They also find that the critical Rayleigh number decreases precipitously with an increase in the depth ratio, η_1, beyond the critical value.

In a subsequent paper, Chen and Chen (1989) report experiments to verify their earlier findings which were in disagreement with the experimental results of Sun (1973). For their experiments they use a rectangular test section with 3 mm diameter glass beads for the porous layer and water, 60 and 90 % glycerin-water solutions, and 100 % glycerin as the saturating fluids. The depth ratio η_1 is varied from 0–1. Fluids of increasing viscosity are used for cases with larger η_1 so as to keep the temperature difference across the test section within reasonable limits. The experiments show a precipitous decrease in the critical Rayleigh number as the depth of the fluid layer is increased from zero and an eight-fold decrease in the critical wavelength between $\eta_1 = 0.1$ and 0.2. These findings confirm their earlier theoretical predictions. Convective cells are observed to be three-dimensional via flow visualization.

Vasseur, Wang and Sen (1989) investigate the onset of convection in a composite system that is heated at the bottom by a uniform heat flux. Fluid flow in the porous layer is modeled using the Darcy-Brinkman equation and the continuity of velocity and shear stress is applied at the interface. To obtain an approximate analytical solution to the problem, they consider a shallow enclosure for which $A \to \infty$. In addition, they assume that for this limiting case the velocity in the central portion of the domain is parallel and is only directed in the vertical direction.

Solutions are obtained for the case where the upper surface of the fluid layer is either rigid or free. For the case of the rigid upper surface, the critical porous layer Rayleigh number is $Ra_{mc}^* = 720Da[\eta\kappa + (1 - \eta)]/(1 - \eta)^5$.

Chen (1990) reports a linear stability analysis in the two-dimensional composite system with vertical through-flow. He finds that both stabilizing and destabilizing factors can be enhanced due to vertical through-flow, thus allowing more precise control of the buoyancy-driven instability in either the fluid or the porous layer. For $\eta_1 = 0.1$, the onset of convection occurs in both fluid and porous layers, the relation between the critical Rayleigh number for the porous layer, Ra_{mc}, and the through-flow strength, ζ, is linear, and the effect of the Prandtl number is insignificant. For $\eta_1 > 0.2$, the onset of convection is largely confined to the fluid layer, and the relation becomes $Ra_{mc} \sim \zeta^2$, except for $Pr = 1$, where $Ra_{mc} \sim \zeta^3$.

Chen et al. (1991) then consider the onset of convection in a system consisting of a fluid sub-layer over a porous sub-layer with anisotropic permeability and thermal diffusivity. Flow in the porous medium is assumed to be governed by Darcy's law and the Beavers-Joseph condition is applied at the interface between the two layers. The effects of anisotropy on convection onset are found to be most profound for small values of the depth ratio, η_1. For fixed values of the vertical permeability, decreasing the value of the horizontal-to-vertical permeability ratio, ξ, leads to stabilization because of increased resistance to motion in the porous sub-layer. At larger values of η_1, the onset of motion is increasingly confined to the fluid layer, with the transport of heat through the porous layer being primarily by conduction. The authors conclude that the influence of ξ on the stability characteristics for larger η_1 is less significant than the effects of anisotropic thermal conductivity.

Chen and Hsu (1991) extend this work to cover a wide range of depth ratios, horizontal-to-vertical permeability ratios, ξ, and horizontal-to-vertical thermal diffusivity ratios, χ. For $\eta_1 < 0.1$, the critical Rayleigh number, Ra_{mc}, is an explicit function of χ/ξ, and the corresponding critical wave number, a_{mc}, is a function of $(\xi\chi)^{1/4}$. For $\eta_1 < 0.1$, the porous sub-layer dominates the system by convection, and anisotropic and inhomogeneous effects are significant as demonstrated by a well-defined dependence of the critical Rayleigh number on the permeability and diffusivity ratios. For $\eta_1 > 0.1$, however, the authors do not obtain any explicit function for Ra_{mc} or a_{mc} in terms of ξ and χ. They note that because the onset of convection is largely confined to the fluid sub-layer anisotropy and inhomogeneity in the porous layer are not significant.

Chen and Lu (1992) investigate the effect of the fluid viscosity on the onset of thermal convection. They use Darcy's law to model fluid flow in the porous sub-layer and apply the Beavers-Joseph boundary condition at the interface. The viscosity variation is represented by the parameter $\Xi = \log(\nu_{max}/\nu_{min})$. They find that the stability characteristics (Ra_{mc}, a_{mc}, and flow structure) are profoundly influenced by the viscosity variation. Intrinsic features of the critical flow are mainly determined by the values of Ξ and η_1. The authors also identify three critical flow patterns on the basis of varying Ξ and η_1 and find that the transition between any two of them is a bimodal instability.

Chen and Chen (1992) further perform a computational analysis to determine the heat transfer characteristics of the two-layer system analyzed in their previous studies. For the porous sub-layer, they consider the Darcy-Brinkman-Forchheimer form of the momentum equation to account for viscous and inertial effects. Boundary conditions at the interface are the continuity of velocity, temperature, heat flux, normal stress, and shear. The flow is assumed to be two-dimensional and periodic in the horizontal direction with a wavelength equal to the critical value at convection onset. They find that convection remains steady for $Ra_m \leq 20\, Ra_{mc}$ when the depth ratio, η_1, is varied between 0.1 and 1.0. For $\eta_1 < 0.13$ (the critical height ratio), the Nusselt number increases sharply with Ra_m, whereas at larger η_1 the increase is moderate. Heat transfer coefficients predicted by the numerical scheme for $\eta_1 = 0.1$ and 0.2 show good agreement with their experimental results.

Prasad and co-workers report a series of experimental studies to visualize flow patterns and measure heat transfer characteristics of a fluid superposed porous medium uniformly heated from below (Prasad and Tian 1990; Prasad et al. 1991; Prasad 1993). In one set of experiments they use 12.7 and 25.4 mm diameter acrylic balls with $0 < \eta < 1$. An immersion method is used to measure the refractive index of the acrylic spheres and match it with the refractive index of silicone oil which is used as the saturating fluid. Aluminum particles are used for visualizing the flow patterns and show that flow channels through large voids produce highly asymmetric and complicated flow structures. The number of convective rolls in the fluid sub-layer is found to depend on η. The heat transfer coefficient generally increases with the Rayleigh number, but its dependence on the non-dimensional particle size, γ, and height ratio, η, is found to be very complex. Generally, average Nusselt numbers increase with γ, but it is found that there may exist some values of η for which the heat transfer coefficient for a smaller γ is larger. Also, the Nusselt number first decreases with an increase in γ and reaches a minimum at γ_{min}. Any further increase in porous layer height beyond this minimum augments the heat transfer rate, and the Nusselt number curves show peaks at $\gamma_{min} < \gamma \leq 1$. A general correlation for the Nusselt number is obtained in the form $Nu = Constant \times Ra^n$, where the constants are found to depend on η and γ.

In another set of experiments, Prasad et al. (1991) use 6, 15, and 24 mm diameter glass beads saturated with distilled water and vary the height ratio in the range $0.067 \leq \eta \leq 1$. Extensive visualization studies using aluminum particles reveal a highly complex, three-dimensional flow field and active flow interactions between the overlying fluid and the packed bed beginning with $Ra_m \sim O(1)$. Strong jet like vertical convective flows are seen to move from the porous sub-layer to the upper fluid layer and results in a sharp drop in the critical Rayleigh number for the porous layer from $4\pi^2$ when $\eta < 1$. The observed flow patterns qualitatively match those predicted by Poulikakos et al. (1986). For small bead diameters, the Nusselt number decreases with increase in η, and for larger bead diameters, it first decreases with η until a γ_{min} is reached and then increases again. For the largest beads, a couple of inflexions on the Nu-versus-γ curve are seen.

Prasad (1993) extends these studies further to include the influence of Prandtl number and different solid–fluid conductivity ratios. Experiments are conducted with acrylic, aluminum, and glass spheres for the porous bed and two kinds of silicone oils and ethylene glycol as the saturating fluid. He finds that the complex relation between Nusselt number and porous layer height remains unaltered with a variation in thermal conductivity ratio and/or the Prandtl number. An increase in thermal conductivity of the solid matrix enhances the heat transfer coefficient. Further, the composite fluid and porous layer can transport more energy than the fluid layer provided the porous matrix is highly permeable and the solid-to-fluid conductivity ratio is large. The effect of Prandtl number follows the trend reported for a cavity completely filled with a porous medium, particularly for large Darcy and Rayleigh numbers.

Kazmeirczak and Muley (1994) measure steady and transient heat transfer characteristics of fully porous and fluid-superposed porous layers. They use 3 mm diameter glass beads as the porous layer and de-ionized water as the saturating fluid. For the steady-state experiments the lower wall is maintained at a constant temperature while for the transient experiments the temperature of the lower wall is varied cyclically. In both cases the temperature of the top wall is constant. With $\eta \approx 0.96$, their results show that the presence of even a very small layer of fluid above the porous layer significantly increases the heat transfer rate when compared to the fully porous layer. This effect is seen for both steady and transient boundary conditions. Heat transfer coefficients in steady convection are correlated by $Nu_m = Ra_m^{0.8145}/13.02$.

Kim and Choi (1996) numerically solve the stability problem for the onset of convection in an overlying fluid-porous layer composite system with the aim of validating the boundary conditions at the interface. Like Poulikakos (1986), they use a single set of conservation equations to model fluid flow and heat transfer in the composite domain. Their scheme is able to accurately predict the number and the wavelength of circulating cells formed at the onset point. They find that when $\eta_1 > 0.12$, the number of circulating cells increases continuously as the Rayleigh number increases, which in turn increases the Nusselt number continuously. However for $\eta_1 < 0.1$, the cellular flows constantly adjust their position and size as the Rayleigh number increases. They also confirm the abrupt and steep drop in the critical Rayleigh number with an increase in the height ratio, η_1, beyond the critical value as observed by Chen and Chen (1992). The number of cells in the supercritical convection regime does not increase monotonically with the Rayleigh number and the corresponding Nusselt number variation is quite different.

Zhao and Chen (2001) re-examine the problem of the onset of thermal and thermo-solutal convection in the two-layer system studied earlier by Chen and Chen (1988). They use a one-domain model to describe fluid flow and heat transfer, and the linear eigenvalue problem is solved using a finite-difference method. Results show that the overall trend of the change in the critical Rayleigh number and wave number with the depth ratio is similar for the two models, but predicted values of the critical parameters vary by ~30–40 % between the two models, e.g., the value of the critical height ratio predicted by the one-domain model is 0.095 as opposed to 0.13 predicted by the two-equation model. Streamlines and isotherms for the one-domain model show convective motion throughout the entire domain as opposed to that of the two-domain model which predicts fluid motion primarily confined to the overlying fluid layer.

Straughan (2002) investigates the effect of fluid and porous layer property variation on the onset of convection. He uses Darcy's law to model flow in the porous layer and applies the Beavers-Joseph boundary condition, and the generalized form proposed by Jones (1973) at the fluid-porous layer interface. The properties varied are the bulk porosity, the Beavers-Joseph constant, Darcy number, effective heat capacity ratio, and effective conductivity ratio. He finds that the critical Rayleigh number at a fixed height ratio decreases with a decrease in Darcy number and an increase the Beavers-Joseph constant.

Steven (2006) reports an experimental study of the heat transfer characteristics of a fluid superposed porous layer system heated from below. His experiments use a cylindrical test chamber with 6 mm diameter glass beads as the porous layer and water as the saturating fluid. Heat transfer coefficients are measured for four values of the height ratio, η, and temperature fluctuations are recorded within the system at various radial and vertical locations. He finds that for a given Rayleigh number, the overall Nusselt number is larger than that for a fully porous layer, indicating the overlying layer enhances the overall heat transfer coefficient. However, the overall Nusselt number does not change significantly with η for different Rayleigh numbers.

Hirata and co-workers (2006, 2007a, b, 2009) reexamine the problem of convective stability in superposed fluid and porous layers. Their aim is to investigate how the modeling of fluid flow in the porous region and the boundary conditions at the fluid-porous layer interface affect the prediction of the onset of convection. Hirata et al. (2006) explore the linear stability problem using a one-domain model using the approach of Zhao and Chen (2001). They find that the marginal stability curves exhibit the same bimodal character as that predicted by the two-domain model. They also obtain excellent agreement with the results of Zhao and Chen (2001) thus confirming the validity of the one-domain model.

In a subsequent study, Hirata et al. (2007a) examine the onset of convective instability via three different modeling approaches: a one-domain approach wherein the porous sub-layer is treated as a pseudo fluid and the entire composite system is modeled with a single set of conservation equations, a two-domain approach wherein flow in the porous layer is modeled by Darcy's law with the Beavers-Joseph condition at the interface, and a modified two-domain approach wherein flow in the porous sub-layer is modeled with the Brinkman extension to Darcy's law and continuity of velocity, temperature, heat flux, normal stress, and shear stress at the interface. The resulting eigenvalue problem in each case is solved with a generalized integral transform technique. They find that marginal stability curves obtained with the two-domain and modified two-domain approaches are in close agreement but differ significantly from the curves obtained with the one-domain approach, indicating that the mathematical formulation of the problem has great influence on the stability results. Their results also show that the effect of the including the Brinkman term in the momentum equation is minimal.

In an extension of this study Hirata et al. (2007b) use a two-domain approach with a different set of interface conditions. Instead of considering the continuity of shear stress at the interface, a stress-jump boundary condition (Ochoa-Tapia and Whitaker 1995a, b) is used with continuity of velocity, temperature, heat flux, and normal stress. They find that the stress-jump coefficient strongly influences the bimodal marginal stability curves. For small fluid-to-porous layer height ratios, increasing the stress-jump coefficient causes the convection in the fluid layer to become unstable. Convection in the porous sub-layer, however, remains unaffected by the magnitude of the stress-jump coefficient, and because convection in the fluid layer occurs due to perturbation of large wave numbers, the stress-jump condition induces a more unstable situation at large wave numbers.

Hirata et al. (2009) repeat their earlier study on stability analysis with the one-domain, two-domain, and modified two-domain approaches to examine the cause for the differences in the marginal stability curves obtained with the one-domain and the two-domain approaches. The conservation equations and the boundary conditions for the three approaches are the same as that used in their earlier studies. However, in their problem formulation for the one-domain approach, they incorporate the hypothesis of Kataoka (1986) that the average properties of the porous medium (porosity, permeability, and effective diffusivity) are Heaviside step functions, and hence their differentiation must be considered in the meaning of distributions. Using this approach they find that marginal stability curves for the one-domain and two-domain models are almost identical and show almost the same bimodal behavior irrespective of the depth ratio. Based on their results, the one-and two-domain approaches are identical provided that the one-domain approach is properly interpreted mathematically, i.e., in the meaning of distributions.

2.2 Conclusion

From the foregoing, it is evident that the ratio of the fluid layer height to the porous layer height exerts a significant influence on the flow and temperature profiles in both the fluid and the porous layer. It also has a significant influence on overall heat transfer rate through the composite system. Numerical studies have shown that the boundary conditions at the fluid-porous layer interface have a great effect on the prediction of convective stability in such a system. Visualization studies of the flow patterns have also shown that there is significant interaction between the overlying fluid and the porous sub-layer. Interface effects are also likely to play a significant role in determining flow patterns, temperature distributions and heat transfer rates when only a fraction of the base of such a composite system is heated. Hence, careful attention must be paid in choosing the governing equations and boundary conditions during problem formulation.

It is clear that there now exists a significant body of literature pertaining to the problem of natural convection in fluid-superposed porous layers heated from

below. These studies examine various fundamental aspects of the problem and
have enabled a thorough understanding of the principal governing criteria. In particular they have identified how parameters, such as the solid-to-fluid conductivity
ratio, fluid Prandtl number, and fluid-to-porous layer height ratio affect convective
heat transfer. The majority of them consider the case where the heat source is uniformly distributed along the bottom of the system, i.e., a fully heated bottom. The
few studies that consider a localized bottom heat source deal almost exclusively
with saturated porous layers ($\eta = 1$). A localized bottom heat source is often a
more realistic thermal boundary condition, and there are apparently no published
studies of this class of problems.

Chapter 3
Mathematical Formulation and Numerical Solution

Keywords Numerical solution • Volume averaging • Interfacial conditions • Two-domain formulation • One-domain formulation • Finite-volume method

In this chapter, the mathematical formulation of the convective heat transfer problem is presented. The governing equations for natural convection in two-dimensional fluid and saturated porous layers are described first. Thereafter the boundary conditions for the problem are elucidated. In particular, boundary conditions at the interface of the fluid and porous layers are discussed in detail. The one-domain formulation is then derived, and the governing equations are presented in dimensionless form using the vorticity-stream function formulation.

3.1 Governing Equations

The fluid-porous layer system is shown schematically in Fig. 3.1. A horizontal fluid layer of thickness H_f extends over a saturated porous layer of thickness H_m. The two-layer system is confined to a two-dimensional enclosure of overall height H and length L, and is bounded on all four sides by impermeable boundaries. The two vertical walls are adiabatic, the upper horizontal wall is held at a constant temperature, T_C, and the lower boundary has a centrally heated portion of length L_H that is held a constant temperature, T_H, while the remaining portion is kept adiabatic. The system is potentially unstable, i.e. $T_H > T_C$, causing a buoyancy driven flow instability.

Two sets of continuity, momentum and energy equations describe natural convection in the system. For the fluid layer,

$$\nabla \times \mathbf{u}_f = 0, \tag{3.1}$$

A. Bagchi and F. A. Kulacki, *Natural Convection in Superposed Fluid-Porous Layers*, SpringerBriefs in Thermal Engineering and Applied Sciences, DOI: 10.1007/978-1-4614-6576-8_3, © The Author(s) 2014

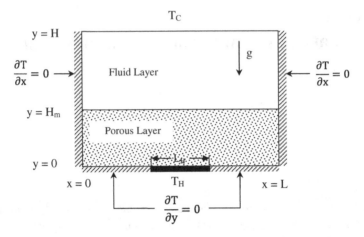

Fig. 3.1 Solution domain of superposed system. $T_H > T_C$. $\delta < 1$

$$\rho \left[\frac{\partial \mathbf{u}_f}{\partial t} + (\mathbf{u}_f \times \nabla)\mathbf{u}_f \right] = -\nabla P_f + \mu_f \nabla^2 \mathbf{u}_f + \beta\rho \left(T_f - T_0 \right) \mathbf{g}, \qquad (3.2)$$

$$\frac{\partial T_f}{\partial t} + \left(\mathbf{u}_f \times \nabla \right) T_f = \alpha_f \nabla^2 T_f, \qquad (3.3)$$

where $\mathbf{u}_f$ is the fluid velocity vector ($\mathbf{u}_f = u_f \mathbf{i} + v_f \mathbf{j}$), ρ denotes the fluid density, t is time, P_f is fluid pressure, μ_f is the fluid dynamic viscosity, β is the coefficient of thermal expansion of the fluid, T_f is the fluid temperature, T_0 is a reference temperature, and α_f is the fluid thermal diffusivity. The gravitational acceleration vector $\mathbf{g}$ points in the negative y-direction. In accordance with the Boussinesq approximation, the fluid density is assumed to be constant everywhere except in the buoyancy term of the momentum equation where its dependence on temperature is assumed to be linear and is,

$$\rho = \rho_0 \left[1 - \beta \left(T_f - T_0 \right) \right], \qquad (3.4)$$

where ρ_0 denotes the fluid density at T_0. This linear assumption is valid when the temperature difference across the domain is not very large, as in the present problem. Hoewever, in many cases a non-linear relation between the density and temperature may be more appropriate. For example, in problems where water is the saturating fluid and the problem formulation involves the density maximum of water at $T_0 = 4$ °C, a quadratic relation such as $\rho = \rho_0[1 - (\beta/2) \left(T_f - T_0 \right)^2]$ may be more appropriate (Eklund 1963).

The corresponding continuity, momentum and energy equations for the porous sub-layer are,

$$\nabla \times \mathbf{u}_m = 0, \tag{3.5}$$

$$\rho \left[\frac{\partial \mathbf{u}_m}{\partial t} + (\mathbf{u}_m \times \nabla) \frac{\mathbf{u}_m}{\phi} \right] = -\nabla P_m + \mu_{\text{eff}} \nabla^2 \mathbf{u}_m - \left(\frac{\mu \phi}{K} \right) \mathbf{u}_m$$
$$- \left(\frac{\rho \phi F}{\sqrt{K}} \right) |\mathbf{u}_m| \, \mathbf{u}_m + \beta \rho \, (T_m - T_0) \, \mathbf{g}, \tag{3.6}$$

$$\sigma \frac{\partial T_m}{\partial t} + (\mathbf{u}_m \times \nabla) \, T_m = \alpha_m \nabla^2 T_m, \tag{3.7}$$

where, $\mathbf{u}_m$, T_m and P_m represent the volume averaged velocity, temperature, and pressure respectively, ϕ is the bulk porosity, K is the bulk permeability, μ_{eff} is the effective viscosity, and F is the Forchheimer coefficient. The heat capacity ratio is,

$$\sigma = \frac{\phi (\rho c_p)_f + (1 - \phi)(\rho c_p)_{\text{so}}}{(\rho c_p)_f}. \tag{3.8}$$

The effective thermal diffusivity is defined $\alpha_m = k_m/(\rho c_p)_f$, and the effective conductivity k_m is,

$$k_m = \phi k_f + (1 - \phi) \, k_{\text{so}}. \tag{3.9}$$

The values of K and F depend on the structure of the porous medium. For a porous medium comprising randomly packed spheres, for example, the permeability and Forchheimer coefficient are,

$$K = \frac{d^2 \phi^3}{175 (1 - \phi)^2}, \tag{3.10}$$

$$F = \frac{1.75 \phi^{-\frac{3}{2}}}{\sqrt{175}}. \tag{3.11}$$

It is instructive at this point to consider in some detail the form of the governing equations for transport through a porous medium as given above. The continuity, momentum and energy equations for the porous sub-layer are valid at the macroscopic scale and represent the volume averaged forms of these equations over a representative elementary volume (REV). The REV includes both the solid and the fluid components of the porous medium and is chosen in a manner such that it is much larger than the pore volume but much smaller than the volume of the entire porous medium (Fig. 3.2). The volume-averaged equations are useful from an engineering standpoint but at the expense of detailed information concerning the microscopic structure of the porous medium. In particular, they do not account for tortuosity, the nature of the boundaries between the solid and fluid phases, and the actual variation of quantities, such as the pressure, within the pores. The gross

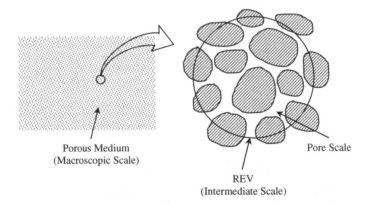

Porous Medium
(Macroscopic Scale)

Pore Scale

REV
(Intermediate Scale)

Fig. 3.2 A representative elementary volume (REV)

macroscopic effects of these factors are retained in the form of parameters, such as the porosity, permeability and effective viscosity, and they can be related to the statistical properties of the porous medium. Thus the volume averaged equations can adequately capture the effects of the microstructure at a macroscopic level without making the problem mathematically intractable. For this reason, the volume averaged governing equations are generally of immense importance in engineering analysis of transport in porous media.

Another important aspect that needs to be discussed is the form of the energy equation. By writing a single energy equation for both the fluid and the solid phases, it is implicit that a condition of local thermal equilibrium exists between the two phases. This is a reasonable assumption when the solid-to-fluid conductivity ratio, λ, is close to unity. For larger conductivity ratios, especially $\lambda \gg 1$, this condition is not strictly true. In that case, separate energy equations are needed for the solid and the fluid phases, and the thermal resistance between the phases needs to be taken into account. This two-equation model however makes the numerical solution much more involved and for this reason, will not be considered here. An equilibrium model such as the present one can nevertheless highlight several important aspects of the problem for large conductivity ratios. As a result, this model will be used to study problems with conductivity ratios as high as $\lambda = 100$, though it must be borne in mind that the results are only for qualitative understanding and may not yield correct quantitative predictions.

3.2 Boundary and Initial Conditions

At the system boundaries,

$$u_f = 0, \; v_f = 0, \; T_f = T_c, \quad at \quad y = H \tag{3.12}$$

$$u_m = 0, \quad v_m = 0, \quad \begin{cases} T_m = T_H, \ |y = 0, \ \frac{L-L_H}{2} \le x \le \frac{L+L_H}{2} \\ \frac{\partial T_m}{\partial y} = 0, \ |y = 0, \quad \text{elsewhere} \end{cases} \tag{3.13}$$

$$u = 0, \quad v = 0, \quad \frac{\partial T}{\partial x} = 0, \quad at \quad x = 0, L. \tag{3.14}$$

The initial conditions for the problem can be written,

$$u = 0, \quad v = 0, \quad T = 0 \quad at \quad t \le 0. \tag{3.15}$$

At the fluid-porous medium interface,

$$T_f = T_m, \tag{3.16}$$

$$k_f \frac{\partial T_f}{\partial y} = k_m \frac{\partial T_m}{\partial y}. \tag{3.17}$$

Unlike the thermal boundary conditions, however, the momentum boundary conditions at the interface do not have a unique formulation. Three different sets of boundary conditions are commonly used and there is as yet no consensus on which formulation accurately captures the fluid dynamics at the interface. The first, proposed by Beavers and Joseph (1967), postulates a discontinuity in the interfacial tangential velocity and shear stress but assumes the normal velocity and stress to be continuous. This is expressed as,

$$v_f = v_m, \tag{3.18}$$

$$-P_f + 2\frac{\partial v_f}{\partial y} = -P_m, \tag{3.19}$$

$$\frac{\partial u_f}{\partial y} = \frac{\hat{\alpha}}{\sqrt{K}} \left(u_f - u_m \right), \tag{3.20}$$

where $\hat{\alpha}$ is the Beavers-Joseph constant. It must be noted that the Beavers-Joseph boundary condition cannot be used with the full Darcy-Brinkman-Forchheimer formulation of the momentum equation in the porous layer (Eq. 3.6) but only in conjunction with the Darcy formulation which is,

$$\rho \frac{\partial \mathbf{u}_m}{\partial t} = -\nabla P_m - \left(\frac{\mu\phi}{K} \right) \mathbf{u}_m + \beta\rho \left(T_m - T_0 \right) \mathbf{g}. \tag{3.21}$$

A second formulation proposed by Neale and Nader (1974) assumes the continuity of velocity, normal stress and tangential shear stress at the interface given as,

$$\mathbf{u}_f = \mathbf{u}_m, \tag{3.22}$$

$$-P_f + 2\frac{\partial v_f}{\partial y} = -P_m + \frac{2}{\phi}\frac{\partial v_m}{\partial y}, \tag{3.23}$$

$$\frac{\partial u_f}{\partial y} = \frac{1}{\phi}\frac{\partial u_m}{\partial y}. \tag{3.24}$$

In addition to the above approaches, a third set of momentum boundary conditions proposed by Ochoa-Tapia and Whitaker (1995a, b) replaces the continuity of shear stress in the Neale-Nader formulation by the stress-jump boundary condition,

$$\frac{\partial u_f}{\partial y} - \frac{1}{\phi}\frac{\partial u_m}{\partial y} = \frac{\hat{\beta}}{\sqrt{K}}(u_{\text{int}}), \tag{3.25}$$

where $\hat{\beta}$ is the stress-jump coefficient and u_{int} is the tangential velocity at the interface.

In spite of these different formulations, several studies have shown that all three sets of boundary conditions lead to near-identical solutions. Singh and Thorpe (1995) find that for buoyancy driven flow in fluid-superposed porous layers heated from the sides, both the Beavers-Joseph and Neale-Nader boundary conditions predict similar flow structures and overall heat transfer rates for Darcy numbers up to 10^{-3}. Similar agreement between the three different formulations was recently demonstrated by Hirata and co-workers (Hirata et al. 2007a, b, 2009). These results suggest that either of these approaches is adequate for the present problem. However the continuity conditions of Neale and Nader have some inherent advantages over the other two formulations. First, the continuity conditions do include an arbitrary empirical parameter like the Beavers-Joseph constant or the stress-jump coefficient. Secondly, unlike the Beavers-Joseph formulation, the continuity conditions can be used in conjunction with the full Darcy-Brinkman-Forchheimer formulation of the momentum equation, thus allowing its application to porous media with a wide range of Darcy numbers. Finally, the continuity conditions have a key advantage in that they allow use of the one-domain formulation in which the two sets of governing equations for the fluid and porous layers can be combined into a single set of equations. In other words, both the porous and the fluid layers can be modeled as a single domain by one set of equations, the solution of which satisfies the continuity of velocity, stress, temperature, and heat flux at the interface. This greatly simplifies the numerical solution procedure especially during writing, debugging and compiling the computer code. Because of these advantages, the continuity boundary conditions of Neale and Nader have been adopted in this study.

3.3 One-Domain Formulation

The one-domain formulation requires a single set of governing equations for the entire composite domain. These equations are presented here using the stream function-vorticity representation for the momentum equation. With this formulation, the pressure term in the momentum equation can be eliminated, which simplifies the numerical solution. To derive the one-domain formulation, we define,

$$x^* = \frac{x}{H}, \quad u^* = \frac{u}{(\alpha/H)}, \quad T^* = \frac{T - T_H}{T_H - T_C}, \tag{3.26a}$$

$$y^* = \frac{y}{H}, \quad v^* = \frac{v}{(\alpha/H)}, \quad t^* = \frac{t}{(H^2/\alpha)}, \tag{3.26b}$$

where the superscript, $(\ )^*$, denotes a dimensionless quantity. The dimensionless stream function and vorticity are thus,

$$\omega^* = \frac{\partial v^*}{\partial x^*} - \frac{\partial u^*}{\partial y^*}, \tag{3.27}$$

$$u^* = -\frac{\partial \psi^*}{\partial y^*}, \quad v^* = \frac{\partial \psi^*}{\partial x^*}, \tag{3.28}$$

and the dimensionless stream function-vorticity equation is,

$$\frac{\partial^2 \psi^*}{\partial x^{*2}} + \frac{\partial^2 \psi^*}{\partial y^{*2}} = \omega^*. \tag{3.29}$$

Note that Eq. (3.29) satisfies the continuity equation identically.

Using the above definitions, the non-dimensional governing equations for the fluid sub-layer are,

$$\frac{\partial \omega^*}{\partial t^*} + (\mathbf{u}^* \times \nabla)\,\omega^* = \Pr\phi \nabla^2 \omega^* + \text{Ra}\Pr\phi^2 \frac{\partial T^*}{\partial x^*}, \tag{3.30}$$

$$\frac{\partial T^*}{\partial t^*} + (\mathbf{u}^* \times \nabla)T^* = \nabla^2 T^*. \tag{3.31}$$

Similarly, the governing equations for the porous sub-layer are,

$$\phi \frac{\partial \omega^*}{\partial t^*} + (\mathbf{u}^* \times \nabla)\,\omega^* = \Pr\phi \nabla^2 \omega^* - \left(\frac{\Pr\phi^2}{\text{Da}} + \frac{F\phi^2}{\sqrt{\text{Da}}} |\mathbf{u}^*| \right) \omega^*$$

$$+ \frac{F\phi^2}{\sqrt{\text{Da}}} \left(u^* \frac{\partial |\mathbf{u}^*|}{\partial y^*} - v^* \frac{\partial |\mathbf{u}^*|}{\partial x^*} \right) + \text{Ra}\Pr\phi^2 \frac{\partial T^*}{\partial x^*}, \tag{3.32}$$

$$\sigma \frac{\partial T^*}{\partial t^*} + (\mathbf{u}^* \times \nabla)T^* = \kappa \nabla^2 T^*. \tag{3.33}$$

To derive the one-domain formulation for the vorticity transport equation, consider the definitions of the permeability and Darcy number. In the fluid sub-layer, porosity has no meaning, i.e. $\phi \equiv 1$, and the permeability and Darcy number are,

$$\phi \to 1 \quad \Rightarrow \quad K \to \infty, \; \text{Da} \to \infty. \tag{3.34}$$

Substituting Eq. (3.34) into Eq. (3.32), the vorticity transport equation for the porous sub-layer reduces to that for the fluid layer in the limit $\phi \to 1$. Thus Eq. (3.32) alone is sufficient to represent vorticity transport in the composite fluid-porous layer system. The porosity acts as a switching parameter that allows the vorticity transport equation to take on the appropriate form depending on the solution domain. This conditionality can be expressed,

$$\phi = \begin{cases} 1 & \text{in the fluid layer} \\ \phi & \text{in the porus layer} \end{cases}. \tag{3.35}$$

The one-domain formulation for the energy equation can be derived in a similar manner. For the fluid sub-layer, the thermal conductivity ratio, stagnant conductivity, and the conductivity ratio can be written,

$$\phi \to 1 \quad \Rightarrow \quad \sigma \to 1, \quad k_m \to k_f \quad \text{and} \quad \kappa \to 1. \tag{3.36}$$

Thus when $\phi \to 1$, the energy equation for the porous sub-layer reduces to the energy equation for the fluid sub-layer. This equation is therefore the one-domain representation of the energy equation.

In summary, the dimensionless governing equations are,

$$\frac{\partial^2 \psi}{\partial x^2} + \frac{\partial^2 \psi}{\partial y^2} = \omega, \tag{3.37}$$

$$\phi \frac{\partial \omega}{\partial t} + (\mathbf{u} \times \nabla)\,\omega = \text{Pr}\phi \nabla^2 \omega - \left(\frac{\text{Pr}\phi^2}{\text{Da}} + \frac{F\phi^2}{\sqrt{\text{Da}}} |\mathbf{u}| \right) \omega$$
$$+ \frac{F\phi^2}{\sqrt{\text{Da}}} \left(u \frac{\partial |\mathbf{u}|}{\partial y} - v \frac{\partial |\mathbf{u}|}{\partial x} \right) + \text{RaPr}\phi^2 \frac{\partial T}{\partial x}, \tag{3.38}$$

$$\sigma \frac{\partial T}{\partial t} + (\mathbf{u} \times \nabla) T = \kappa \nabla^2 T, \tag{3.39}$$

where the superscript, ( )*, has been dropped for convenience. Dimensionless quantities will henceforth be written without any superscripts.

The corresponding dimensionless boundary conditions are,

$$\psi = \frac{\partial \psi}{\partial x} = 0, \quad \omega = \frac{\partial^2 \psi}{\partial y^2}, \quad T = 0, \quad \text{at} \quad y = 1, \tag{3.40}$$

$$\psi = \frac{\partial \psi}{\partial x} = 0, \quad \omega = \frac{\partial^2 \psi}{\partial y^2}, \quad \begin{cases} T = 1, \; |y = 0, \; \frac{L - L_H}{2} \le x \le \frac{L + L_H}{2} \\ \frac{\partial T}{\partial y} = 0, \; |y = 0, \quad \text{elsewhere} \end{cases} \tag{3.41}$$

$$\psi = \frac{\partial \psi}{\partial y} = 0, \quad \omega = \frac{\partial^2 \psi}{\partial x^2}, \quad \frac{\partial T}{\partial x} = 0, \quad \text{at} \quad x = 0, L/H. \tag{3.42}$$

The initial conditions are,

$$u = v = 0, \quad \omega = \psi = 0, \quad T = 0, \quad \text{at} \quad t < 0. \tag{3.43}$$

The mathematical model contained in Eqs. (3.26)–(3.43) comprises a set of coupled partial differential equations which in general cannot be solved analytically. Therefore the governing equations are discretized and solved numerically using a control-volume formulation (Spalding 1972), which ensures that the conservation laws for mass, momentum and energy are satisfied for any group of control volumes, as well as over the entire solution domain. A uniform grid is used, with each control volume containing a grid point at its geometric center, and the grid is constructed in a manner such that no grid point is located along the porous-fluid layer interface. A central differencing scheme is used for all second-order derivatives. Advective terms in the vorticity transport and energy equations are discretized using the QUICK scheme (Leonard 1979, Hayase et al. 1992), and the Darcy, Forchheimer and buoyancy terms are treated as source terms. Linearization of the source terms is performed according to the general recommendations of Patankar (1980), and diffusion coefficients for the control volume faces located along the fluid-porous layer interface are obtained using the harmonic mean formulation. This formulation can handle abrupt changes in diffusion coefficients across control volumes without requiring an excessively fine grid.

The linear discretized equations are solved using a direct method based on the up-looking Cholesky factorization technique. Solution of the coupled set of discretized equations is started by first solving the vorticity transport equation. Thereafter the Poisson equation for the stream function and the energy equation are solved. This procedure is repeated iteratively until steady-state is attained under the convergence criterion,

$$\frac{\sum_{i=1}^{m} \sum_{j=1}^{n} \left| \Theta_{i,j}^{r+1} - \Theta_{i,j}^{r} \right|}{\sum_{i=1}^{m} \sum_{j=1}^{n} \left| \Theta_{i,j}^{r+1} \right|} < 10^{-5}, \tag{3.45}$$

where Θ denotes ω, ψ, or T, and r denotes the iteration number. Under relaxation of the stream function equation assures convergence of the iterative solution, and the value of the under relaxation parameter is 0.8. Grid fineness required to obtain a convergent solution depends strongly on Ra, η, and A. In general, a finer grid is required at high Rayleigh number and near the critical Rayleigh number. Finer grids are also needed to obtain convergent solutions at large values of η and A. The dimensionless time step is $\Delta t = 0.1$, and the solution is somewhat insensitive to the value of the time step because of the fully implicit time integration scheme.

Table 3.1 Grid convergence of the numerical solution. $Ra = 10^5$, $\eta = 0.5$, $\delta = 0.5$

Grid	Nu	% change	Energy balance (%)
20 × 40	4.5064	–	0.82
40 × 80	4.3765	2.8813	2.63
52 × 104	4.3302	1.0583	2.44
64 × 128	4.2950	0.8135	2.32
80 × 160	4.2579	0.8635	2.17

Table 3.1 summarizes grid convergence behavior. Further details are given by Bagchi (2010).

Heat transfer results are expressed in dimensionless form using the average conduction-referenced Nusselt number over the heated surface,

$$Nu = \frac{qH}{k_{\text{eff}}L_H (T_H - T_C)} = \frac{k_m}{Ak_{\text{eff}}} \int_0^A \left(\frac{\partial T}{\partial y}\right)_{y=0} dx, \qquad (3.46)$$

where k_{eff} is the effective conductivity of the composite domain defined as,

$$k_{\text{eff}} = k_f \left[(1 - \eta) + \frac{\eta}{\phi (1 - \phi) \lambda}\right]^{-1}. \qquad (3.47)$$

A convergence test of the numerical scheme is carried out by evaluating the average Nusselt number for successively finer grids and calculating the percentage change with grid refinement. The solution is considered to be grid independent when any further change in grid size produces less than a one percent change in the Nusselt number. As an additional test on the accuracy of the numerical solution, an energy balance is computed,

$$\Delta E = \left[\frac{(\text{Total Heat Input}) - (\text{Total Heat Output})}{(\text{Total Heat Input})}\right] = \left(\frac{Nu - Nu_T}{Nu}\right), \qquad (3.48)$$

where the Nusselt number along the top surface, Nu_T, is

$$Nu_T = \frac{qH}{k_{\text{eff}}L (T_H - T_C)} = \frac{k_f}{Ak_{\text{eff}}} \int_0^A \left(\frac{\partial T}{\partial y}\right)_{y=1} dx. \qquad (3.49)$$

Because of the integral energy conservation of the finite volume method, an exact solution would yield a zero value for the energy balance. However, due to the accumulation of numerical and round-off errors during the solution, the value of the energy balance is not zero but a very small number. For all simulations discussed below, the energy balance is satisfied to <2.5 percent.

3.4 Verification of the Numerical Scheme

To verify the accuracy of the numerical solution technique, simulations are performed with $\phi = 0.39$, $F = 0.5$, $\eta = 0.5$ and $\lambda = 1$. These parameter values are suitable for a randomly packed layer of glass beads saturated with water and have been selected to facilitate the comparison of the numerical results with experiments. The overall heat transfer coefficient is the quantity of engineering significance, and verification of the solution method is obtained by comparing overall Nusselt numbers to accepted values for four classical porous media heat transfer problems. First, results are obtained for the Rayleigh-Bénard problem, which is a special case of superposed layer system with $\eta = 0$ and $\delta = 1$. For this simulation, $A = 4$ and $Pr = 0.71$ (air). The critical Rayleigh number is found to be ~1920, six percent higher than the critical value of 1,810 obtained with linear stability theory for a domain with $A = 4$ (Platten and Legros 1984). Beyond the critical Rayleigh number, excellent agreement is obtained with the results of Soong et al. (1996, Fig. 3.3a) whose problem parameters are identical to ours. Our results also agree with the numerical results of Clever and Busse (1974) for an infinitely wide fluid

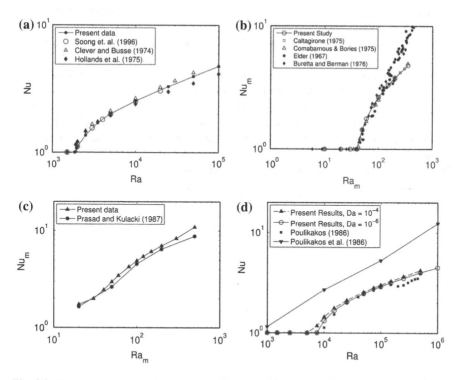

Fig. 3.3 Comparison of numerical results to literature data for **a** the Rayleigh-Bénard problem, **b** the Horton-Rogers-Lapwood problem, **c** the Elder problem, and **d** the uniformly heated fluid-superposed porous layer problem

layer, as well as with the experimental results of various investigators compiled by
Hollands et al. (1975).

The second test problem is the Horton-Rogers-Lapwood problem which is
a special case of the present problem for $\eta = 1$ and $\delta = 1$. Figure 3.3b shows a
comparison of the present results with numerical results of Caltagirone (1975)
and Combarnous and Bories (1975). Excellent agreement is seen, and the critical
Rayleigh number for the onset of convection predicted via the numerical scheme is
40, very close to the theoretical value of 39.48. It is well known that for convection
in saturated porous layers, there exists a second critical point ($Ra_m \sim 380$) beyond
which oscillatory convection prevails, and no steady-state is attained. Here no
attempt is made to evaluate this second critical Rayleigh number, although it must
be noted that no unique steady solution is obtained for $Ra_m > 400$. Figure 3.3b also
shows excellent agreement of the present results with the experiments of Elder
(1967) and Buretta and Berman (1976) except at very high Ra_m. This trend has
been reported in earlier numerical studies (Combarnous and Bories 1975).

Comparison of the present results to those of Prasad and Kulacki (1987) for the
Elder problem is shown in Fig. 3.3c. Again, excellent agreement is obtained and
the maximum difference between the two data sets is ~ 7 % except at $Ra_m = 500$
where the Nusselt number predicted by the present calculations is about 15 %
higher. The disagreement is most likely due to inertial effects which become significant at high Rayleigh number. Because Prasad and Kulacki use Darcy's law
to model fluid motion in the porous layer, effects of fluid inertia which are now
accounted for are not captured.

Comparison of results for the uniformly heated fluid-superposed porous layer
problem with literature data is shown in Fig. 3.3d. Results for $Da = 10^{-4}$ and
10^{-6} are compared to those of Poulikakos (1986) and Poulikakos et al. (1986).
Excellent agreement is obtained with the results of Poulikakos but differ significantly from those of Poulikakos et al. This disagreement arises most likely
because while Poulikakos uses a one-domain model, Poulikakos et al. use a two-
domain model with the Beavers-Joseph condition at the interface. The disagreement between the different numerical results is an important issue to be discussed
in Chap. 6.

Chapter 4
Numerical Prediction of Convection

Keywords Local heating natural convection • Heater length ratio • Fluid-porous height ratio • Darcy number • Prandtl number • Aspect ratio • Conductivity ratio

The governing conservation equations contain seven dimensionless parameters that determine the heat transfer characteristics of the superposed fluid-porous layer system. They are the heater-to-base length ratio, δ, the porous layer-to-total height ratio, η, the overall aspect ratio, A, the Darcy number, Da, the fluid Prandtl number, Pr, the solid-to-fluid conductivity ratio, λ, and the overall Rayleigh number, Ra. To present a thorough parametric analysis of the problem, the effect of each parameter on the flow and temperature fields and on overall heat transfer coefficients is discussed in this chapter. This exposition of results will aid in identifying the most important parameters and developing a general heat transfer correlation that is applicable to a wide range of problems where this system is encountered. The values of the parameters used in the calculations are given in Table 4.1.

4.1 Effect of Heater Size

To clearly understand the effects of having a localized heat source at the base, it is important first to examine the flow and temperature fields. Figures 4.1, 4.2, and 4.3 show steady streamlines and isotherms for $\eta = 0.5$ and $\delta = 1$, 0.5 and 0.25, respectively at four Rayleigh numbers. For $\delta = 1$ (a uniformly heated base), there is no convective motion at Ra $= 10^3$, and the system is in the conduction mode. Although streamlines show a cellular convective motion, these are meaningless as the gradient of the stream function is extremely low. When the Rayleigh number increases to 10^4, convective motion commences and two pairs of counter rotating cells are seen. The corresponding isotherms show a rising thermal plume. Convective motion is confined to the overlying fluid layer with some penetrative convection into the underlying porous layer. Penetrative convection in composite

A. Bagchi and F. A. Kulacki, *Natural Convection in Superposed Fluid-Porous Layers*,
SpringerBriefs in Thermal Engineering and Applied Sciences,
DOI: 10.1007/978-1-4614-6576-8_4, © The Author(s) 2014

Table 4.1 Parameters used for the numerical solution

Parameter	Range	Fixed parameters	Grid size
Ra	$Ra = 10^3\text{–}10^6$	$\delta = 0.5$ $\eta = 0.5$ $A = 2$ $Da = 10^{-6}$	$40 \times 80\ (Ra < 10^4)$ $52 \times 104\ (10^4 < Ra < 10^6)$ $64 \times 128\ (Ra = 10^6)$
η	$\eta = 0.25, 0.5, 0.75$	$Ra = 10^5$ $\delta = 0.5$ $A = 2$ $Da = 10^{-6}$	$\underline{\eta = 0.25, 0.5:}$ $40 \times 80\ (Ra < 10^4)$ $52 \times 104\ (10^4 < Ra < 10^6)$ $64 \times 128\ (Ra = 10^6)$ $\underline{\eta = 0.75:}$ $40 \times 80\ (Ra < 10^4)$ $80 \times 160\ (10^4 < Ra \leq 10^6)$
δ	$\delta = 0.25, 0.5, 0.75$	$Ra = 10^5$ $\eta = 0.5$ $A = 2$ $Da = 10^{-6}$	$40 \times 80\ (Ra < 10^4)$ $52 \times 104\ (10^4 < Ra < 10^4)$ $64 \times 128\ (Ra = 10^6)$
A	$A = 2, 4, 6$	$Ra = 10^5$ $\eta = 0.5$ $\delta = 0.5$ $Da = 10^{-6}$	$\underline{A = 2, 4:}$ $40 \times 80\ (Ra < 10^4)$ $52 \times 104\ (10^4 < Ra < 10^6)$ $64 \times 128\ (Ra = 10^6)$ $\underline{A = 6:}$ $40 \times 80\ (Ra < 10^4)$ $64 \times 128\ (10^4 < Ra \leq 10^6)$
Da	$Da = 10^{-2}, 10^{-4}, 10^{-6}$	$\delta = 0.5$ $\eta = 0.5$ $A = 2$ $Ra = 10^5$	$40 \times 80\ (Ra < 10^4)$ $52 \times 104\ (10^4 < Ra < 10^6)$ $64 \times 128\ (Ra = 10^6)$

systems has been the subject of a large number of numerical studies in the literature but no definitive conclusion can be drawn from the evidence available. Some studies report the occurrence of penetrative convection (Poulikakos 1986; Poulikakos et al. 1986) while others (Chen and Chen 1992; Kim and Choi 1996) report that convection is limited to only the upper fluid sub-layer with minimal or no penetration into the porous sub-layer. This lack of agreement cannot be simply attributed to the modeling of the interfacial boundary conditions given that studies using a particular boundary condition have reported different results.

Visualization studies by Prasad and co-workers show that penetrative convection does indeed occur (Prasad and Tian 1990; Prasad et al. 1991; Prasad 1993). However, due to the large pore diameters of the medium used in their studies, generality cannot be ascribed to this conclusion. It is sufficient to say that the present results are in accordance with the mathematical formulation of the problem which assumes the continuity of velocity components across the interface, and hence implicitly allows for penetrative convection. Thus, the present results indicate that beyond the critical Rayleigh number, a state of conduction with no fluid motion in

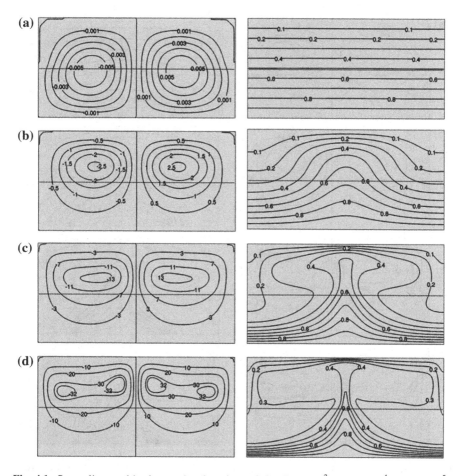

Fig. 4.1 Streamlines and isotherms. $\delta = 1$, and $\eta = 0.5$. **a** $Ra = 10^3$, **b** $Ra = 10^4$, **c** $Ra = 10^5$, and **d** $Ra = 10^6$. The *dashed horizontal line* indicates the location of the fluid-porous layer interface

the underlying porous layer never exists. It must be noted that convective motion in the porous sub-layer occurs primarily in the vicinity of the interface. In the region away from the interface, there is little convective motion and heat transfer is mainly by conduction.

With an increase in the Rayleigh number to 10^5, the velocities of convective motion in the overlying fluid layer increase, though the extent of flow penetration into the porous sub-layer does not increase significantly. Individual cellular flows appear to be horizontally stretched, and isotherm patterns show a narrow plume rising along the centerline of the cavity. At a $Ra = 10^6$, the plume becomes narrower and two separate pockets of re-circulating flow can be seen within the same cell.

When the size of the heater covers only half of the base ($\delta = 0.5$), isotherms change significantly (Fig. 4.2). However, streamlines show little noticeable change

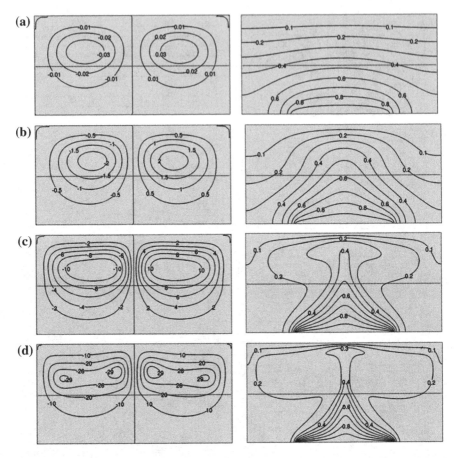

Fig. 4.2 Streamlines and isotherms. $\delta = 0.5$, and $\eta = 0.5$. **a** Ra $= 10^3$, **b** Ra $= 10^4$, **c** Ra $= 10^5$, and **d** Ra $= 10^6$

except that at Ra $= 10^3$, a cellular convective pattern is seen. Though the gradient of the stream function at this Rayleigh number is small, it is not negligible which indicates that a circulatory fluid motion with low velocity exists. Fluid motion does not, however, significantly enhance the overall heat transfer rate, and isotherms are essentially identical to those obtained for conduction. Thus, convection is not the dominant mode of heat transfer and cellular fluid motion cannot be attributed to the onset of flow. Circulatory fluid motion arises due to a horizontal temperature gradient at the edges of the localized heat source which initiates fluid motion. Such end cells have been observed in convection in porous and fluid layers with localized heat sources. It must be noted that the end cells are not located at the heater edges, but in the overlying fluid layer. This phenomenon can be explained by recalling that for convection in fluid-superposed porous layers, fluid motion is always confined to the overlying fluid layer except for large height

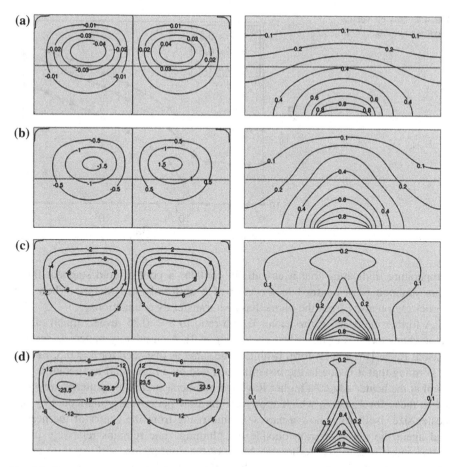

Fig. 4.3 Streamlines and isotherms. $\delta = 0.25$, and $\eta = 0.5$. **a** Ra $= 10^3$, **b** Ra $= 10^4$, **c** Ra $= 10^5$, and **d** Ra $= 10^6$

ratio ($\eta > 0.91$). Thus, whether the source of instability is either latent in the system or imposed by a local heat source, the mode of convection is essentially the same. The heat source, therefore, acts as a trigger for the onset of motion and the ensuing flow patterns remain unaffected by whether heating at the base is uniform or localized.

With increasing Rayleigh number, the overall flow structure remains essentially the same as that for a uniform heat source. Flow remains confined to the fluid layer, and the extent of penetration into the porous sub-layer does not increase. The isotherms, however, differ significantly from those for a fully heated bottom. With increasing Ra, the plume rising above the heat source is seen to become narrower. However, the porous sub-layer outside the plume and away from the interface remains essentially isothermal and does not participate significantly in energy transport. The temperature in this region is very close to the upper surface

Fig. 4.4 Nusselt-versus-
Rayleigh number. $\eta = 0.5$

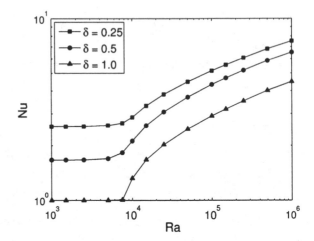

Fig. 4.4 Nusselt-versus-
Rayleigh number. $\eta = 0.5$

temperature, indicating that it essentially contains a pool of cold stagnant fluid. This situation is in contrast to the case of a uniformly heated base where the porous region away from the plume has heat transfer by conduction.

A further decrease in the heater length ratio to $\delta = 0.25$ reveals much of the same characteristics as described above (Fig. 4.3). A circulatory pattern is seen to exist at Ra = 10^3, and its form is almost the same as that seen for $\delta = 0.5$, further confirming that it is due to the instability created by a horizontal temperature gradient at the heater edge. At higher Rayleigh numbers, streamline patterns are similar to those for $\delta = 0.5$, which show that the flow patterns are unaffected by the heater size. Isotherms show a narrow plume rising from the center of the heater, and again the porous region outside the pluming zone remains relatively unaffected and is essentially isothermal. Much of the fluid region, and therefore much of the composite domain, is essentially isothermal at Ra = 10^6. This is likely to become more pronounced as the heater size shrinks further until, in the limiting case of a point heat source, the region outside the rising plume is likely to be completely quiescent.

The effect of the localized heat source on the overall heat transfer coefficient is shown in Fig. 4.4. For a uniformly heated base, the critical Rayleigh number for the onset of convection is ~7.5 $\times$ 10^3, below which the average Nusselt number is unity indicating that heat transfer occurs solely by conduction. When the heater size is finite, the Nusselt number is no longer unity even at low values of Ra, a consequence of the fact that the presence of a localized heat source initiates a circulatory motion due to the temperature gradient at the heater edge, as mentioned above. Hence, a pure conduction state never exists for a localized heat source. This, however, does not imply the dominant mode of heat transfer is convection. Even for discrete heat sources, the Nusselt number remains approximately constant until the critical point is reached. While the pre-critical point heat transfer regime is not purely a conduction regime, it is conduction dominated. Also noteworthy is that while the transition to the convective regime is discrete for $\delta = 1$, the transition to the convective regime is now a smooth one. This aspect has been previously observed for convection in fully porous and fluid layers with discrete heat sources and is a consequence of the existence of a base flow before convection becomes the dominant energy transport mechanism. As such, there is no sudden initiation of convective motion but a gradual strengthening of the existing base flow pattern.

Over the entire range of Rayleigh number, the average Nusselt number increases with a decrease in the size of the heat source. This has been previously observed with respect to convection in porous and fluid layers and can be understood by recalling back the discussion on the flow and isotherm patterns. As mentioned earlier, for a uniformly heated base, heat transfer away from the pluming region is primarily by conduction, which for a low solid-to-fluid conductivity ratio ($\lambda = 1$ here) is fairly small. On the other hand, for $\delta < 1$, most of the heater area falls within the pluming region, and almost all the energy input is carried away by convection. In this case, conduction occurs only near the edges of the heated section. Therefore with a decrease of heater size, more of the energy input is transferred to the top via the rising thermal plume and leads to higher Nusselt numbers for smaller heater lengths.

To conclude the discussion on the effects of a localized heat source, the temperature profiles along the centerline of the cavity at Ra = 10^5 are shown in Fig. 4.5. It can be seen that with decrease in the size of the heated fraction of the base, the temperature profiles along the centerline of the plume change significantly. For example, the temperature at the interface ($y/H = 0.5$) decreases with a decrease in the heater length, which holds for temperatures throughout the entire cavity height. With decreasing heater size, much of the temperature drop along the plume occurs very close to the heater while in the rest of the plume, the temperature changes gradually. With decreasing heater size, the plume draws away more and more of the heat that is supplied to the system.

Fig. 4.5 Dimensionless temperature on the vertical centerline. $\eta = 0.5$, and Ra $= 5 \times 10^5$

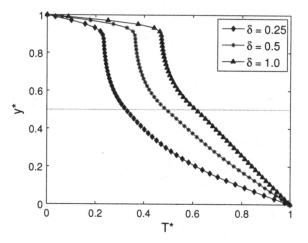

4.2 Effect of Porous Sub-Layer Height

Similar to the approach adopted earlier, the description of the effect of the porous-
to-total height ratio begins with a discussion of the flow and temperature fields at
different height ratios. In these simulations, all parameters are kept fixed except
for the height ratio. Figures 4.2, 4.6, and 4.7 show streamlines and isotherms for
$\eta = 0.5$, 0.25 and 0.75, respectively, and $\delta = 0.5$, $A = 2$, $Da = 10^{-6}$, $Pr = 7$, and
$\lambda = 1$.

First, the case of $\eta = 0.5$ will be discussed as this case (Fig. 4.1) has already
been encountered in the previous section. To briefly summarize the previous discussion, the salient feature is that the convective motion is restricted to the fluid
sub-layer with some penetration into the porous sub-layer. A circulatory motion
triggered by the presence of a localized heat source is seen at $Ra \leq 10^3$, although

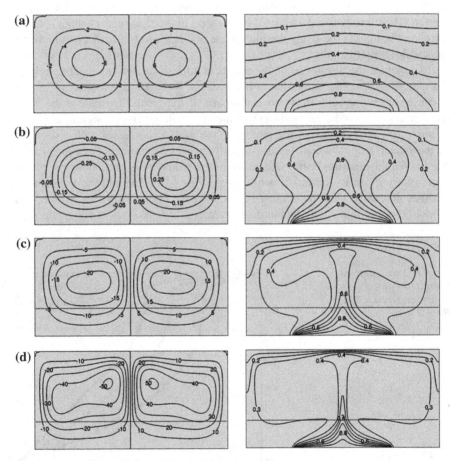

Fig. 4.6 Streamlines and isotherms. $\delta = 0.5$ and $\eta = 0.25$. **a** $Ra = 10^3$, **b** $Ra = 10^4$, **c** $Ra = 10^5$,
and **d** $Ra = 10^6$

convection is yet not the dominant mode of energy transport. These two observations point to the fact that conduction heat transfer never exists in the porous sub-layer when a localized heat source is present. Thus, a sharp critical point for the onset of convection cannot be defined for a localized heat source; a critical point in this case merely indicates that convection becomes the dominant mode of heat transfer. The isotherms for $\eta = 0.5$ indicate that with increasing Ra, the plume rising from the central portion of the heater becomes narrower and the region outside the plume in the porous layer is essentially isothermal.

When the height ratio drops to $\eta = 0.25$ (Fig. 4.6), streamlines and isotherms are very much similar to those obtained for fluid convection with a localized heat source. With three-quarters of the entire domain occupied by the fluid layer, convection exists in almost the entire cavity. At Ra $= 10^3$, circulatory motion driven by the localized heat source can be seen. A reduction in volume of porous matrix

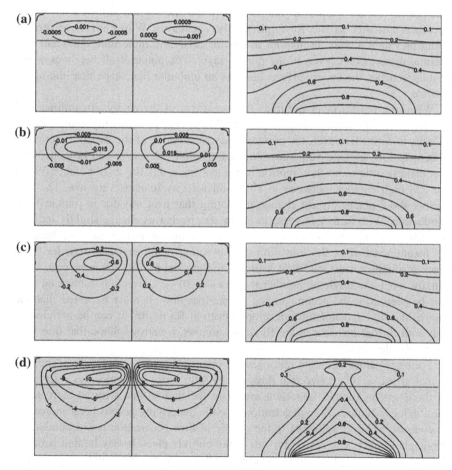

Fig. 4.7 Streamlines and isotherms. $\delta = 0.5$ and $\eta = 0.75$. **a** Ra $= 10^3$, **b** Ra $= 10^4$, **c** Ra $= 10^5$, and **d** Ra $= 10^6$

in the domain reduces overall resistance to fluid motion and leads to higher fluid velocities. This trend is seen for all values of Ra. Based on this observation, it can be deduced that the overall heat transfer coefficient increases with a decrease in the height ratio due to a reduction in the resistance to fluid motion by the porous layer and an associated increase in convection. With an increase in the Rayleigh number, the velocity of fluid motion increases. The shape of the cellular motion, however, is different from that for $\eta = 0.5$. The cellular shape is almost square and noticeably larger as a direct consequence of the convection being dominated by the overlying fluid layer. As fluid motion is restricted primarily to the fluid layer, the size, and shape of the circulating cells adjust to fit the fluid space. With the aspect ratio fixed, the width of the cells is thus constrained. At Ra $= 10^6$, there is an indication that the flow may switch from bicellular to quadricellular. Unfortunately, the limits of computation do not permit a verification of this trend at larger Rayleigh numbers. Isotherms are similar to those observed for $\eta = 0.5$. The central feature is a plume-like flow along the mid-section of the cavity, which becomes narrower as the Rayleigh number increases. With convective motion being prevalent almost throughout the entire cavity, no noticeable quiescent isothermal region exists, even in the porous layer. The plume itself becomes wider at it approaches the upper surface, and has an umbrella like shape near the upper surface.

When the porous layer occupies three-quarters of the cavity, streamlines and isotherms change significantly (Fig. 4.7). At Ra $= 10^3$, instability causes a pair of cells to form. However, the maximum absolute value of the stream function is extremely low, making velocities almost negligible. Hence, it can be concluded that there is little or no fluid motion in the system at this Rayleigh number. When the Rayleigh number increases to 10^4, fluid velocity remains very low. The isotherms show practically no change indicating that heat transfer is primarily by conduction. With further increase in Ra to 10^5, isotherms change slightly indicating that convection begins to take over as the dominant heat transfer mechanism. As mentioned earlier, the transition to convection dominated heat transfer takes place gradually for discrete heat sources, and gradual transition can be seen here. Finally, when the Rayleigh number increases to 10^6, a convection dominated flow can be seen with the accompanying plume-like flow. It must be noted that the plume-like flow has not fully developed even at Ra $= 10^6$. It can be anticipated that with further increase in the Rayleigh number, a narrow plume-like flow will eventually develop.

For large values of height ratio, several important aspects can be seen. The first is that at Ra $= 10^6$, convective flow occurs with fairly low velocities over most of the domain. Another important aspect of the flow is that with an increase in the Rayleigh number, flow penetration into the underlying porous layer increases. In contrast, at lower values of η, the degree of flow penetration is independent of Rayleigh number. The reason for this is not entirely clear. It may be that penetration takes place simply to accommodate the flow pattern which does not have sufficient space within the fluid layer to develop. An important aspect that must be noted is that because of the values of the height ratio chosen no bimodal behavior,

Fig. 4.8 Nusselt-versus-
Rayleigh number. $\delta = 0.5$

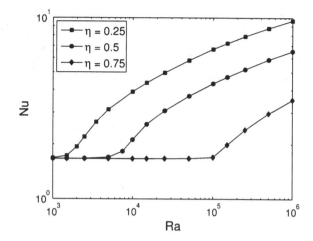

indicated by a sudden change in the value of the critical Rayleigh number, is
observed. The value of the critical height ratio at which this sudden change occur
is ≈ 0.89 (Chen and Chen 1988, 1989), which is well beyond the maximum height
ratio examined here.

Further insight into the effects of the flow field can be gained from the overall
heat transfer coefficient. Figure 4.8 shows Nusselt-versus-Rayleigh number with
height ratio as a parameter. The results mirror the inferences that can be drawn
from an analysis of the flow and temperature fields. The onset point for convection dominated heat transfer increases with the increase in the fraction of the
cavity occupied by the porous layer given that $Ra_m \sim Ra\eta^3\kappa$. It has been shown
using linear stability theory that with an increase in η, the critical value of Ra_m
increases (Sun 1973; Chen and Chen 1988). This implies that with an increase in
the height of the porous layer, much larger values of Ra are needed to induce convective motion of higher intensities. However, the Nusselt number in the conduction regime remains the same at all height ratios and is expected as the heater size
remains fixed.

At Rayleigh numbers larger than the critical value, overall heat transfer coefficients are larger for lower height ratios. This is a direct consequence of the fact
that convective motion for a given Rayleigh number is stronger when the value of
η is lower, i.e., when the thickness of the overlying fluid layer is higher. This trend
has also been noticed in numerical and experimental studies (Poulikakos 1986;
Chen and Chen 1992; Prasad and Tian 1990; Prasad et al. 1991). Finally, it must
be noted that the Nusselt number curves all have the same shape irrespective of
the height ratio. A close look at the curves will show that they are almost parallel
indicating a common mechanism of heat transfer with a different point of origin
that is dictated by the height ratio. This can be further confirmed by looking at the
dimensionless temperature profiles along the centerline of the cavity as shown in
Fig. 4.9. It can be seen that the temperature profiles for all height ratios converge

Fig. 4.9 Dimensionless
centerline temperature.
$\delta = 0.5$, $Ra = 5 \times 10^5$. The
three horizontal lines indicate
the locations of the fluid–
porous interface for different
height ratios

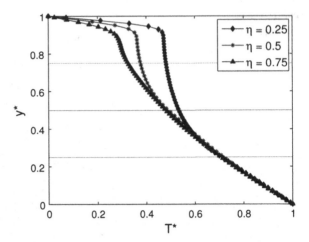

to the same value within the porous layer and the difference among them is noticeable only in the fluid layer. Within the fluid layer, centerline temperatures increase
with the height ratio leading to higher heat flux at the upper boundary.

4.3 Porous Structure Effects

The Darcy number is a dimensionless measure of the permeability of the porous
medium and it is of vital importance in understanding how heat transfer coefficients
change with the permeability of the system. It must be noted that the Darcy number
in the present study is defined as $Da = K/H^2$, where H is the overall height of the
composite system. This definition of the Darcy number thus depends on the system
under consideration and does not express the intrinsic Darcy number of the porous
sub-layer which would be based on a length scale suitable for it. However, a Darcy
number based on a different length scale can be derived directly from the current
definition of Da. For example, a Darcy number based on the pore diameter, d, can be
written $Da_p = \gamma^2 Da$. Conclusions based on the current definition of the Darcy number are therefore applicable to alternative definitions of the Darcy number.

Consider solutions for $Da = 10^{-6}$, 10^{-4}, and 10^{-2}. Figures 4.2, 4.10, and 4.11
show the streamline and isotherm patterns at four Rayleigh numbers. Comparing
the flow patterns for $Da = 10^{-4}$ and 10^{-6} it can be seen that there is not much difference between them. At $Ra = 10^3$, localized heater driven convection patterns
can be seen. Fluid velocities for $Da = 10^{-4}$ are slightly higher than those for 10^{-6}
as indicated by the slight difference in the value of the stream function. This trend
is expected as a porous medium with a higher Darcy number offers lower resistance to fluid motion, and hence allows for higher flow velocities. Isotherms, however, show that heat transfer is yet conduction dominated at this value of Ra.

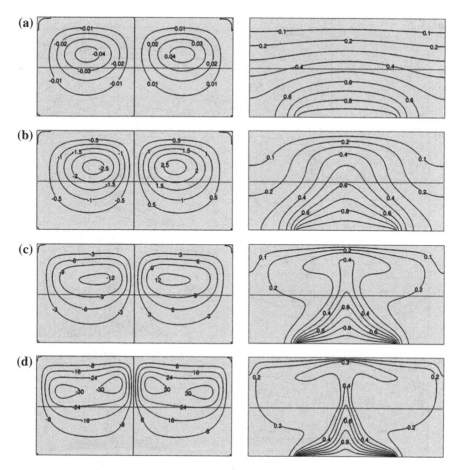

Fig. 4.10 Streamlines and isotherms. Da $= 10^{-4}$, $\delta = 0.5$ and $\eta = 0.5$. **a** Ra $= 10^3$, **b** Ra $= 10^4$, **c** Ra $= 10^5$, and **d** Ra $= 10^6$

With an increase in Rayleigh number to 10^4, convection dominated flow can be seen for both values of the Darcy number. Fluid velocities increase and a pluming pattern can be seen to start forming in both cases. The values of the stream function indicate that flow velocities are almost identical in both cases. This is also the case when the Rayleigh number is increased to 10^5. The flow and temperature fields are virtually identical indicating that an increase in the Darcy number from 10^{-6} to 10^{-4} has little effect on the overall flow field, and hence heat transfer coefficients. When the Rayleigh number increases to 10^6, however, there is a noticeable difference between the two cases. The streamlines show that while flow velocities are comparable, there is increased penetration of flow into the underlying porous layer for Da $= 10^{-4}$. Isotherms also show that temperatures in the fluid sub-layer increase with an increase in Darcy number. Significant differences in flow patterns, and as a consequence overall heat transfer coefficients, between these two cases can be seen only at very high Rayleigh numbers.

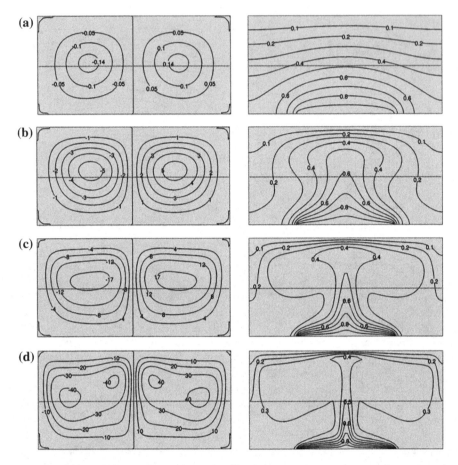

Fig. 4.11 Streamlines and isotherms. Da $= 10^{-2}$, $\delta = 0.5$ and $\eta = 0.5$. **a** Ra $= 10^3$, **b** Ra $= 10^4$, **c** Ra $= 10^5$, and **d** Ra $= 10^6$

When the Da $= 10^{-2}$, significant changes can be seen. Most importantly, the flow is now no longer confined to the upper fluid layer but is instead of spread out across the two layers, a direct consequence of the increased permeability of the porous layer which allows much higher levels of flow penetration. The higher value of the Darcy number also means that the porous sub-layer offers minimal resistance to convective flow which leads to higher flow velocities. With an increase in Rayleigh number, the intensity of the convective motion increases, and the convective cells acquire an almost square shape, indicating that convective motion in the fluid and porous layers is comparable. Flow patterns at Ra $= 10^6$ bear a great resemblance to those seen in Rayleigh-Bénard convection. Further insight can be gained by looking at the isotherms. At Ra $= 10^3$, isotherms show that the system is yet in the conduction mode. At Ra $= 10^4$, a plume-like flow is

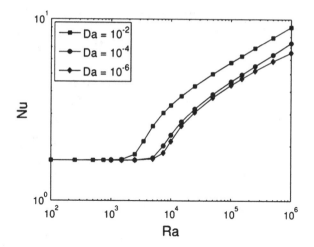

Fig. 4.12 Nusselt versus Rayleigh number relation. $\delta = 0.5$, and $\eta = 0.5$

seen which is much more developed as compared to the isotherms for lower Darcy number. Interestingly, the thermal plume has an almost uniform character throughout the height of the cavity, whereas at lower Darcy numbers, the plume had different characteristics in the fluid and porous regions. Thus, it can be deduced that the temperature profile along the plume will be significantly different from the profiles for lower Darcy numbers.

Overall heat transfer results for different Darcy numbers are shown in Fig. 4.12. The Darcy number has an effect on both the onset of convection and on heat transfer coefficients beyond the critical point. As the Darcy number increases from 10^{-6} to 10^{-4}, the critical Rayleigh number decreases slightly, although this is not very clearly visible. With further increase in Da, however, the critical Rayleigh number drops noticeably. The Nusselt number in the conduction dominated heat transfer regime, however, is the same for all Darcy numbers as the heater length is kept fixed. Once convection starts dominating, Nusselt numbers for Da $= 10^{-6}$ and 10^{-4} are very close to each other except at very high Rayleigh numbers where the effects of lower permeability can be seen. It is likely that this effect is due to the increased effect of inertia. Heat transfer coefficients for Da $= 10^{-2}$ are, however, much higher than those at lower Darcy numbers. This result is expected based on the prior discussion of flow patterns which shows that high levels of convective penetration into, and reduced resistance to fluid motion by, the porous sub-layer lead to increased heat transfer coefficients. Dimensionless temperature profiles along the centerline (Fig. 4.13) also indicate that the temperature distribution for Da $= 10^{-2}$ is significantly different from the temperatures at lower Darcy numbers. This is especially true for temperature within the porous sub-layer and clearly shows that at such high values of the Da it has an almost fluid-like character. These results may be important in applications where the porous layer is either loosely packed or a material with a foam-like microstructure.

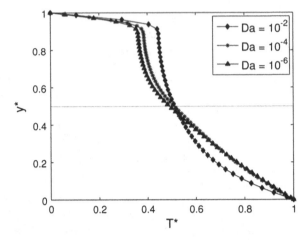

Fig. 4.13 Dimensionless centerline temperature. $\delta = 0.5$, $\eta = 0.5$ and Ra $= 5 \times 10^5$

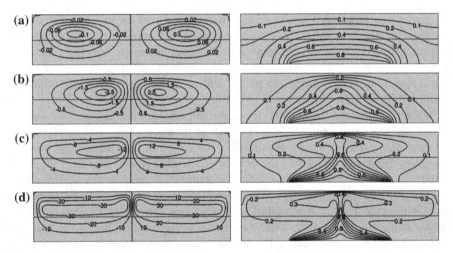

Fig. 4.14 Streamlines and isotherms. $A = 4$, $\delta = 0.5$, and $\eta = 0.5$. **a** Ra $= 10^3$, **b** Ra $= 10^4$, **c** Ra $= 10^5$, and **d** Ra $= 10^6$

4.4 Effect of Aspect Ratio

Aspect ratio is a very important parameter in determining the overall heat transfer coefficient, especially when the heat source is localized. Prior investigations of convection in fluid and porous layers with localized heat sources have shown that there is a complex relation between aspect ratio and the heater length (Prasad and Kulacki 1987; Rajen and Kulacki 1987; Papanicolau and Gopalakrishna 1995). It is, therefore, worthwhile to investigate whether such a relation exists for convection in fluid superposed porous layers. Figures 4.2, 4.14, 4.15, and

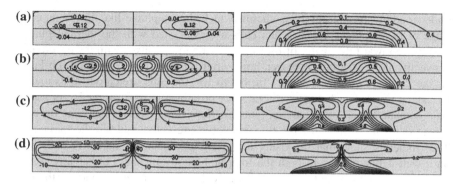

Fig. 4.15 Streamlines and isotherms. $A = 6$, $\delta = 0.5$ and $\eta = 0.5$. **a** Ra $= 10^3$, **b** Ra $= 10^4$, **c** Ra $= 10^5$, and **d** Ra $= 10^6$

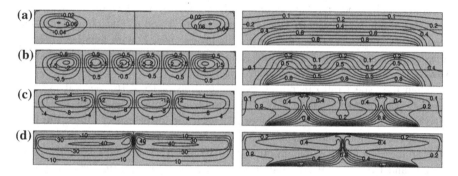

Fig. 4.16 Streamlines and isotherms. $A = 6$, $\delta = 0.75$ and $\eta = 0.5$. **a** Ra $= 10^3$, **b** Ra $= 10^4$, **c** Ra $= 10^5$, and **d** Ra $= 10^6$

4.16 show streamlines and isotherms at four Rayleigh numbers for $A = 2$, 4, and 6. The parameters held constant are $\eta = 0.5$, Da $= 10^{-6}$, $\lambda = 1$, and Pr $= 7$. In Figs. 4.2a and 4.14a, it can be seen that the aspect ratio does not have any significant influence on the flow and temperature fields at Ra $= 10^3$. The circulating flow triggered by the heater edge temperature gradient is seen for both aspect ratios. Interestingly, the cells for $A = 4$ are far apart from each other and are almost centered on the vertical lines drawn at the edges of the heater, further indicating their origin point. This feature cannot be seen at lower aspect ratios due to the smaller horizontal extent of the domain. The velocity of convective motion is also slightly higher at $A = 4$. The corresponding isotherms indicate that convection dominated heat transfer has not yet commenced in either system.

At Ra $= 10^4$, convection dominated heat transfer can be seen for both aspect ratios. It can be seen that the shape and the position of the circulating cells changes once convective flow is underway. The cells move away from their original positions and toward each other. This change in position can be better understood by considering the isotherms which show a plume-like flow in an early stage of development. Because the cells are associated with the thermal plume, they move toward the center of the heater to align themselves with the developing convective flow. The flow, however, largely remains confined to the fluid sub-layer. As the Rayleigh number increases further to 10^5, the cells stretch outwards toward the lateral walls. This behavior can be more prominently seen for the larger aspect ratio domain. Values of the stream function, however, are nearly the same for both aspect ratios indicating that the flow velocities are comparable in the two cases. The isotherms show that the rising thermal plume becomes narrower and takes on a well-defined shape in both the long and short domains. Further increase in the Rayleigh number causes the circulating cells to stretch further and the thermal plume to become narrower. An interesting aspect of the flow is that the number of cells does not increase with an increase in the aspect ratio, in contrast to the case of a uniformly heated base where the wavelength, and hence the number of the cells is determined by the aspect ratio. Thus for $A = 4$, there would be four circulating cells when the base is uniformly heated. However when the heating is localized, the number of cells is a function of both A and δ. This dependence has been observed also in convection in porous layers ($\eta = 1$) with localized heating (Prasad and Kulacki 1987; Rajen and Kulacki 1987). In the present case, values of δ and A are such that only two cells are observed.

When the aspect ratio increases to $A = 6$ (Figs. 4.15, 4.16), several interesting flow patterns are seen. At Ra $= 10^3$, circulatory cells triggered by the heater edge temperature gradient can be seen. As seen for lower aspect ratios, the flow is conduction dominated which can be clearly seen from the isotherm patterns. With an increase in the Rayleigh number to 10^4, convection dominated flow commences and two pairs of circulating cells are seen. This is a further example of how the number of cells depends on both the heater length and the aspect ratio. Within each pair of cells, there is, however, a conspicuous lack of symmetry, and the cells toward the center of the domain are smaller while those near the lateral walls are larger. This is primarily a consequence of the readjustment of the size and position of the cells due to the particular combination of the heater length and aspect ratio. With further increase in the Rayleigh number to 10^5, the same flow pattern is seen with higher flow velocities. Cells near the lateral walls lead to flow restructuring, and at Ra $= 10^6$, a single pair remains. These cells span the entire horizontal extent of the cavity, although the largest fluid velocities are in the overlying fluid layer. Isotherms show that with an increase in Rayleigh number, the number of plumes decreases progressively. These results show that when the heat source at the base is localized, the number of cells is a function of the aspect ratio, heater length ratio, and the Rayleigh number. This flow restructuring is better illustrated in Fig. 4.16 which shows a domain with $A = 6$ and $\delta = 0.75$. As the Rayleigh number increases from 10^3 to 10^6, the number of convection cells decreases from six to two while the number of thermal plumes decreases from three to one. Poulikakos (1986) has reported the occurrence of flow restructuring at high Rayleigh numbers in fluid-superposed porous layers heated uniformly. However, no such phenomenon is observed here for $\delta = 1$.

Fig. 4.17 Nusselt versus
Rayleigh number. $\delta = 0.5$,
and $\eta = 0.5$

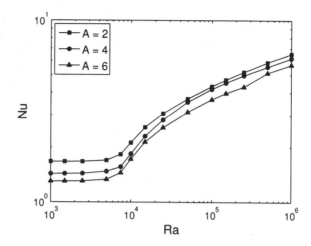

Fig. 4.17 Nusselt versus Rayleigh number. $\delta = 0.5$, and $\eta = 0.5$

The Nusselt-vs-Rayleigh number relation for different aspect ratios is shown in Fig. 4.17. As can be seen, the curves for all aspect ratios have the same pattern across the entire range of Rayleigh numbers, although the overall heat transfer coefficient decreases with an increase in the aspect ratio. In each case, the critical Rayleigh number for the onset of convection dominated heat transfer is approximately the same. The curves also show several interesting features which are a direct consequence of the flow patterns discussed earlier. For $A = 2$ and 4, it can be seen that beyond the onset of convection dominated flow, Nusselt numbers for $A = 4$ increase at a higher rate than that for $A = 2$ until Ra $\approx 5 \times 10^4$, after which they remain approximately parallel indicating that the overall heat transfer coefficient increases at approximately the same rate for both cases. This can be explained by considering the flow patterns in the two systems. At low Rayleigh numbers, a fraction of the domain for $A = 4$ does not have any fluid motion, while for $A = 2$ fluid motion covers almost the entire horizontal extent of the domain. Hence, the system with a lower aspect ratio is able to more efficiently transfer heat by convection. However, with an increase in Ra, the flow spreads further toward the side walls with the larger aspect ratio, and the stagnant fluid near the walls now participates in convection. For smaller aspect ratio domains, however, the flow cannot further spread laterally as it has already reached its maximum extent. As such, the heat transfer coefficient for $A = 4$ increases more rapidly. Once the flow has spread to its maximum lateral extent for the longer system, the rate of increase of the Nusselt number becomes identical for both the aspect ratios. For $A = 6$, the curve is almost parallel to that of $A = 2$ until Ra $\approx 2.5 \times 10^5$. Above this Rayleigh number, there is an abrupt change in the slope of the curve, and Nusselt numbers beyond this point are much closer to those for lower aspect ratios—a direct consequence of flow restructuring that creates a single pair of cells. When the number of cells is larger, there are pockets of stagnant fluid between the cells which do not participate in convection. However, when the cells merge, all of the fluid within the domain participates in convective motion leading to higher overall rates of heat transfer.

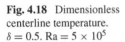
Fig. 4.18 Dimensionless
centerline temperature.
$\delta = 0.5$. Ra $= 5 \times 10^5$

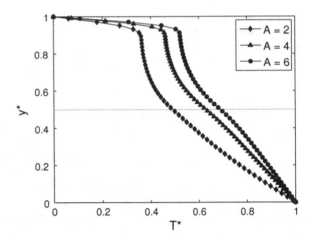

Dimensionless temperature profiles on the centerline of the cavity at Ra $= 10^5$ are shown in Fig. 4.18. It can be seen that along the cavity height, centerline temperatures increase with the aspect ratio. This pattern is remarkably similar to the centerline temperature profiles for different values of the heater length ratio δ, shown earlier in Fig. 4.5. It can be seen that for $A = 2$, the temperature drops much more rapidly within the porous layer than for $A = 4$ and 6, implying that the temperature gradient near the heater increases with a decrease in the aspect ratio. This leads to a more efficient channeling of the energy input and causes the lower aspect ratio domain to have higher heat transfer coefficients.

4.5 Effect of Conductivity Ratio

All simulations discussed thus far have been carried out for the case where the conductivities of the solid and fluid phase are the same, i.e., $\lambda = 1$. In reality, however, this condition is rarely satisfied, and many practical applications involve high solid-to-fluid conductivity ratios. Modeling of convection in porous media with high conductivity ratios, however, poses several challenges. First, a suitable model for the effective stagnant conductivity must be selected as improper representation of this parameter can significantly affect the overall results. Second, for very high conductivity ratios, the assumption of thermal equilibrium between the fluid and solid phases breaks down and, as a result, a single energy equation can no longer be used. Instead, a two-equation model that accounts for the thermal resistance at the solid–fluid interface must be used which makes the numerical solution of the problem highly involved.

To systematically study the effect of the conductivity ratio, it is important to look first at the modeling of the effective conductivity. Aichlmayr and Kulacki (2006) have shown that the effective conductivity data from the various literature sources can be grouped into three categories based on the conductivity ratio: small conductivity ratio ($1 \leq \lambda < 10$), intermediate conductivity ratio ($10 \leq \lambda < 1,000$), and high conductivity ratio ($\lambda \geq 1,000$). They postulate that these ranges demarcate the relative importance of interfacial effects between the solid and fluid phases. High conductivity ratio systems typically have strong interfacial effects and, as a result, require the consideration of separate models for the solid and fluid phases. Due to the difficulty of solving two-equation models, high conductivity ratio systems will not be considered here. Instead, the study will focus on low and intermediate conductivity ratios. Three values of the conductivity ratio are considered here: $\lambda = 1$, 50, and 100. These values have been chosen based on the effective conductivities of glass–water ($\lambda = 1.08$), glass–air ($\lambda = 48$), and steel–water ($\lambda = 102$) systems, all of which are very relevant for practical application. To calculate the effective conductivity, two different models are used. For $\lambda = 1$ the mixture model is used, and for $\lambda = 50$ and $\lambda = 100$, the model of Kunii and Smith (1960) is used.

Streamline and isotherm patterns for $\lambda = 1$, 50, and 100 for different Rayleigh numbers are shown in Figs. 4.2, 4.19, and 4.20 wherein $\eta = 0.5$, $\delta = 0.5$, $A = 2$, Da $= 10^{-6}$, and Pr $= 7$. The case of $\lambda = 1$ has been discussed previously and will be discussed only with respect to the results for the other conductivity ratios. As before, for Ra $= 10^3$, circulation patterns created by the presence of a localized heat source are seen in all cases and the heat transfer is primarily by conduction. Isotherms, however, show significant differences. For $\lambda = 1$, a gradual temperature change occurs across the entire domain. On the other hand, for $\lambda = 50$ and $\lambda = 100$, the temperature gradient across the porous layer is very small and the entire layer is approximately at a constant temperature. The major temperature drop takes place in the fluid layer, a direct consequence of the larger conductivity of the porous sub-layer. Expressed differently, the saturated porous sub-layer has a lower conduction resistance than the overlying fluid layer. As a result, a small temperature difference exists across the porous sub-layer.

When the Rayleigh number is increased to 10^4, convection dominated flow is seen in all three cases. The flow patterns for $\lambda = 1$ appear slightly different than that for $\lambda = 50$ and 100 in terms of the direction of the velocity vectors. Another interesting feature that can be seen is that flow penetration into the underlying porous layer is much higher for $\lambda = 50$ and 100. Isotherm patterns, however, differ significantly for the low ($\lambda = 1$) and high ($\lambda = 50$, 100) conductivity ratios. For $\lambda = 1$ a developing plume-like pattern can be seen at Ra $= 10^4$. However, for $\lambda = 50$ and 100, no clear plume-like flow pattern is visible. Instead, the isotherms, especially those in the fluid layer, indicate that heat transfer is still primarily by conduction. In addition, the isotherms crossing the interface show an abrupt change in slope. This deduction has been observed previously in studies on layered porous media with conductivity contrasts (Lai and Kulacki 1987) and is a direct consequence of the difference in conductivities of the sub-layers.

With further increase in the Rayleigh number to 10^5, a well-defined convective flow is seen in all the three cases. Streamlines indicate that the flow in all the three cases is generally identical. Isotherms patterns show a rising thermal plume

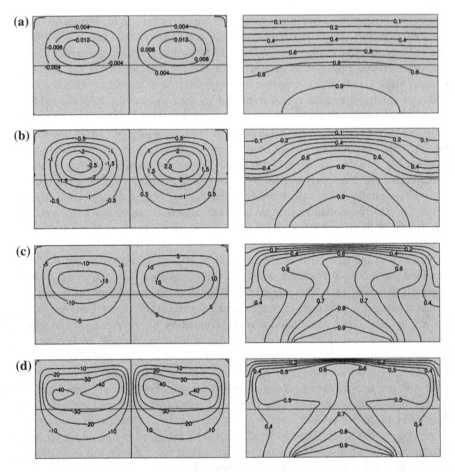

Fig. 4.19 Streamlines and isotherms. $\lambda = 50$, $\delta = 0.5$ and $\eta = 0.5$. **a** Ra $= 10^3$, **b** Ra $= 10^4$, **c** Ra $= 10^5$, and **d** Ra $= 10^6$

for all conductivity ratios, although the plume for $\lambda = 1$ is more developed than that at the higher conductivity ratios. For the higher conductivity ratios, there is a very rapid drop in temperature from the top of the plume to the upper isothermal surface indicating the presence of strong temperature gradients near the upper surface, and thus much higher heat transfer coefficients. When the Rayleigh number increases to 10^6, identical flow patterns are seen in all the three cases. Flow penetration from the fluid layer to the porous sub-layer is higher for the larger conductivity ratios. Isotherms indicate a well-formed rising thermal plume in all cases. As noted earlier, large temperature gradients exist near the top of the plume for the higher conductivity ratios, but the abrupt changes in isotherm gradients at the interface that are seen at lower Rayleigh numbers are much less pronounced at higher Rayleigh numbers. This is another indication that convection is the

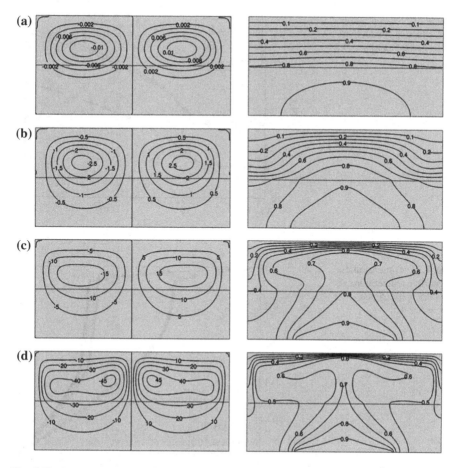

Fig. 4.20 Streamlines and isotherms. $\lambda = 100$, $\delta = 0.5$ and $\eta = 0.5$. **a** Ra $= 10^3$, **b** Ra $= 10^4$, **c** Ra $= 10^5$, and **d** Ra $= 10^6$

dominant mechanism of heat transfer and helps to smooth out the effects of conductivity mismatch between the upper and lower sub-layers.

The Nusselt versus Rayleigh number relation for different conductivity ratios is shown in Fig. 4.21. The figure clearly shows that an increase in the conductivity ratio increases the overall rates of heat transfer through the system as can be expected. A more interesting observation is that the curves for $\lambda = 50$ and 100 are almost similarly indicating that a 2-fold increase in the conductivity ratio does not significantly increase the overall heat transfer coefficient. This can be much better understood by considering the effective conductivity ratio for the two cases. For $\lambda = 50$, $\kappa = 6.1$ whereas for $\lambda = 100$, $\kappa = 7.5$. The effective conductivity is an independent parameter in the governing equations, and the close value of κ is responsible for nearly identical Nusselt versus Rayleigh number curves. Another interesting feature of the graphs is that the curves for $\lambda = 1$, 50, and 100 are not

Fig. 4.21 Nusselt number
and effective conductivity
ratio. $\delta = 0.5$, and $\eta = 0.5$

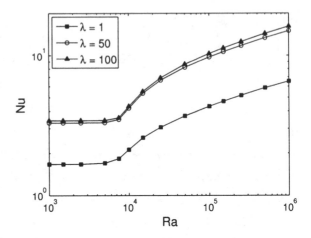
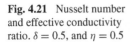

Fig. 4.22 Dimensionless
centerline temperature.
$\delta = 0.5$, $\eta = 0.5$ and
Ra $= 5 \times 10^5$

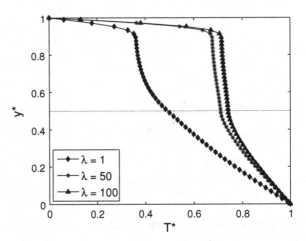

parallel to each other, indicating that it is simply not the higher conductivity of the solid matrix which contributes to the higher heat transfer rates for the higher conductivity ratios.

Temperature profiles along the centerline of the cavity at a Rayleigh number of 5×10^5 are shown for the three conductivity ratios in Fig. 4.22. As can be expected, the profiles for the higher and lower conductivity ratios are widely different. For small values of λ, there is rapid drop in temperature across the porous sub-layer after which the temperature drop in the fluid sub-layer is very low except near the upper surface. However, for higher values of λ, there is a relatively small drop in temperature across the porous sub-layer and practically no drop in the temperature in the fluid sub-layer. Almost all the temperature change occurs near the upper surface where there exists a very large temperature gradient as is seen in previous isotherm patterns. Conduction has a greater influence on the temperature distribution for low conductivity ratios, $1 \leq \lambda \leq 10$. Large values of λ tend to diminish the temperature drop across the porous sub-layer, and the fluid sub-layer exhibits a well-mixed core temperature profile. This inference can also be drawn from isotherm patterns at high Rayleigh number and large conductivity ratio.

4.6 Prandtl Number Effect

The effect of the Prandtl number has been very briefly covered in prior studies (Chen and Chen 1992) of convection in fluid-superposed porous layers heated from below, and thus there is no benchmark with which to compare the present results. As such, wherever possible, results will be compared with prior studies of convection in fluid and porous layers. Two Prandtl numbers are considered here: Pr = 0.7 and 7. These are approximately equal to the Prandtl numbers of air (Pr = 0.71 at 25 °C) and water (Pr = 6.26 at 25 °C), respectively. These fluids have been selected for two primary reasons. First, they are most commonly encountered in laboratory studies of fluid and porous media convection, and there are several references available in the literature for comparison. Also, for these fluids, there is no significant change in the fluid viscosity, and hence the Prandtl number, with temperature. For fluids with higher viscosity, the temperature dependence of Prandtl number cannot be ignored. However, the present model is valid only for constant Pr, and high viscosity fluids, though very important in a large number of applications, are not considered.

Flow and temperature fields for Pr = 7 and 0.7 are shown in Figs. 4.4 and 4.23, respectively. At Ra = 10^3, the flow patterns for both fluids are nearly identical. Two circulating cells produced due to end effects at the heater are seen, and the fluid velocities are identical. Similarly, the isotherms indicate that heat transfer is conduction dominated. With increase in the Rayleigh number to 10^4, flow patterns seen earlier in the chapter (initiation of convection dominated heat transfer, strengthening of the existing cellular flow patterns, and confinement of the flow to the upper fluid layer) are seen for both fluids. Fluid velocities are almost the same in both cases as seen from the value of the stream function. Similarly, isotherms show a gradually developing thermal plume as convection heat transfer rates start increasing.

When the Rayleigh number is increased further to 10^5, however, different streamline patterns are seen. For Pr = 0.7, velocities within the core of the circulation cells are at an angle to the fluid-porous layer interface. On the other hand, for Pr = 7, flow velocities within the core of the cellular flows are parallel to the interface. This effect occurs due to lower shear resistance in the low Prandtl number fluid and is also seen in simulations of Rayleigh-Bénard problem with the fluids of different viscosities. The maximum absolute values of the stream function are almost identical in both cases. Thus velocities, though having different directions within the core, have approximately the same value. Isotherms, however, appear nearly identical indicating that heat transfer rates for the two fluids are

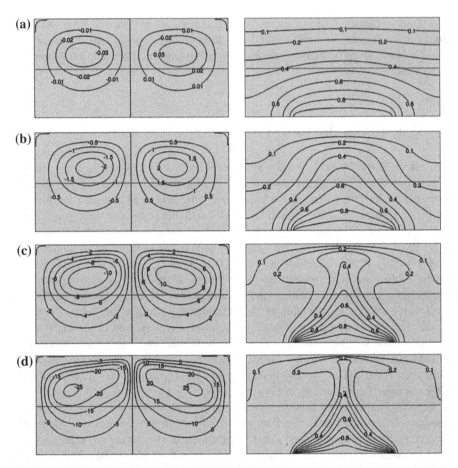

Fig. 4.23 Streamlines and isotherms for Pr $= 0.7$, $\delta = 0.5$ and $\eta = 0.5$. **a** Ra $= 10^3$, **b** Ra $= 10^4$, **c** Ra $= 10^5$, and **d** Ra $= 10^6$

almost same. When the Rayleigh number is increased to 10^6, streamline patterns for Pr $= 0.7$ become highly skewed. The core of the convection cells is seen to become narrow near the rising plume, i.e., near the vertical centerline of the cavity. Isotherms show a narrow rising plume while the region outside the plume is largely isothermal, especially in the porous sub-layer.

The Nusselt versus Rayleigh number relation for the two Prandtl numbers is shown in Fig. 4.24. The curves are almost identical indicating that the Prandtl number has little effect on the overall heat transfer coefficient. This observation is in accordance with the results of both the Rayleigh-Bénard problem and the Horton-Rogers-Lapwood problem. The only noticeable difference in the overall heat transfer coefficients can be seen when Ra $> 10^5$ and can be attributed to the difference in the flow field, e.g., at Ra $= 10^6$. The centerline temperature profiles

Fig. 4.24 Nusselt versus
Rayleigh number. $\eta = 0.5$,
and $\delta = 0.5$

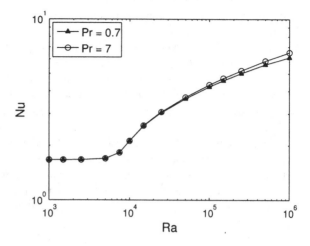

Fig. 4.25 Dimensionless
centerline temperature.
$\delta = 0.5$ and Ra $= 5 \times 10^5$

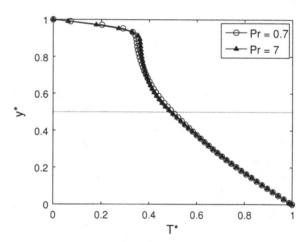

for Ra $= 10^5$ do not show much difference for the two fluids (Fig. 4.25). Thus the
Prandtl number in this range is not a significant parameter in determining overall
heat transfer coefficients.

4.7 Evolution of Temperature Fields

To get a fundamental understanding of the mechanism by which the convective flow
develops from an initial disturbance, we investigate convection with uniform ($\delta = 1$)
and localized ($\delta = 0.5$) heating. For these two simulations, Ra $= 10^5$, $\eta = 0.5$,
Da $= 10^{-6}$, $A = 2$, $\lambda = 1$, and Pr $= 7$. The evolution of the flow and temperature

fields for the two cases is shown in Figs. 4.26 and 4.27 at various values of the dimensionless time starting from $t = 0$ to the final steady state. For $\delta = 1$, the solution is initiated by introducing a small sinusoidal perturbation to the temperature field at $t = 0$ (Fig. 4.26a). At $t = 1$, two convective cells are seen, and flow is almost entirely restricted to the overlying fluid sub-layer. The corresponding isotherms corroborate this observation. The temperature field inside the porous sub-layer resembles a conduction heat transfer mode, but some convective motion can be seen in the fluid layer. At $t = 40$, a much more well-defined flow pattern emerges. Circulatory fluid motion exists in the entire fluid layer and signs of flow penetration into the porous layer appear. The corresponding isotherms exhibit a plume-like flow in an early stage of development. When $t = 60$, an increase in the velocities of convective motion accompanied by further flow penetration into the porous layer occurs. Simultaneously, the thermal plume becomes better defined. As time progresses, this process continues until steady state is attained and flow patterns do not exhibit oscillatory motion. This is consistent with earlier findings on convection in uniformly heated fluid-superposed porous layers (Chen and Chen 1989).

When $\delta = 0.5$, the flow evolution progresses in a slightly different manner. In this case, no initial perturbation is given to initiate the iterative solution. Rather the horizontal temperature gradient at the edge of the heater triggers the onset of convective motion. This can be seen in Fig. 4.27a, where two small end cells are seen at the edge of the heater. Once the onset of convective motion is triggered, a weak circulatory fluid flow is seen in the upper fluid layer at $t = 1$. The corresponding isotherms show only a small disturbance to the conduction temperature field indicating that convective motion has not yet commenced. At $t = 5$, the first signs of the onset of convective motion can be seen. A non-negligible circulatory flow pattern develops in the fluid layer and rising thermal plume begins to take shape. Thereafter, evolution of the flow field follows the same path toward steady state as for $\delta = 1$: fluid velocity increases, there is a flow penetration into the underlying porous layer, and a narrow thermal plume rises from the center of the heater. Thus, the presence of localized heating does not affect either steady-state convection or the path to steady state; rather, it provides a different trigger for the onset of convection.

4.8   Summary

Flow and temperature fields for a wide range of parameters that govern heat transfer coefficients have been determined computationally for two-dimensional laminar flow at Rayleigh numbers up to 10^6. The presence of a localized heat source does not affect the mode of convective motion but provides a different trigger for the onset of convection. Fluid motion is primarily confined to the fluid sub-layer with penetrative convective motion into the porous sub-layer.

Overall heat transfer coefficients increase with decrease in the heater length fraction. A similar trend is noticed with an increase in the Darcy number. Both of

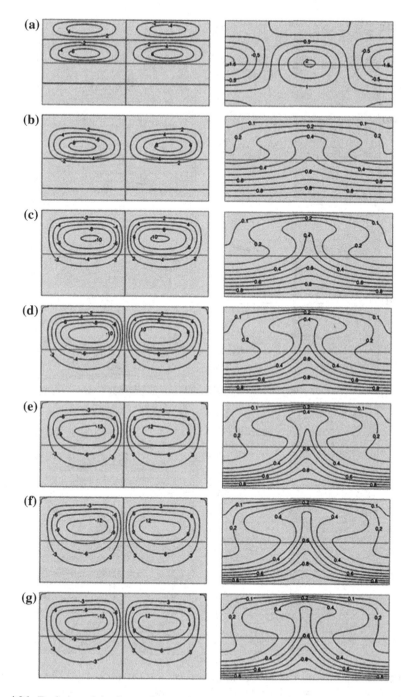

Fig. 4.26 Evolution of the flow and temperature. **a** $t = 0$, **b** $t = 1$, **c** $t = 5$, **d** $t = 10$, **e** $t = 20$, **f** $t = 40$, and **g** $t = 60$. Ra $= 10^5$, $\delta = 1$, $\eta = 0.5$, Pr $= 7$, Da $= 10^{-6}$, $A = 2$, $\lambda = 1$

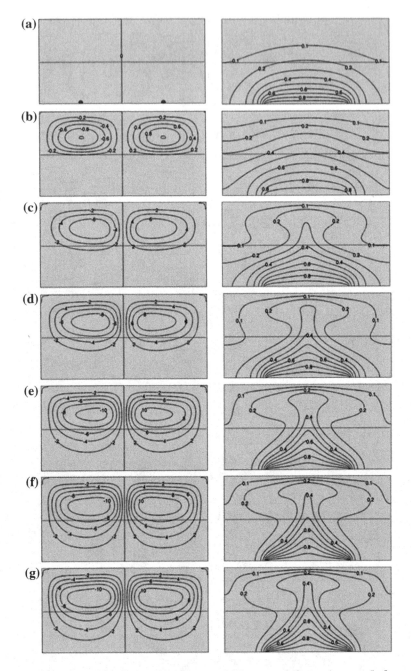

Fig. 4.27 Evolution of the flow and temperature fields. **a** $t = 0$, **b** $t = 1$, **c** $t = 5$, **d** $t = 10$, **e** $t = 20$, **f** $t = 40$, and **g** $t = 60$. $Ra = 10^5$, $\delta = 1$, $\eta = 0.5$, $Pr = 7$, $Da = 10^{-6}$, $A = 2$, $\lambda = 1$

these effects can be attributed to the same fundamental cause: a decrease in the resistance to fluid motion offered by the porous sub-layer. An increase in the solid-to-fluid conductivity ratio leads to significant enhancement of the overall heat transfer coefficient. This is not only due to higher conductivity of the solid matrix but also due to an increase in the intensity of the convective motion with increase in the conductivity ratio. The Prandtl number over the range $0.7 < Pr < 7$ has very little effect on the overall heat transfer coefficient except at very high Rayleigh numbers.

The effect of the aspect ratio on the flow structure and heat transfer coefficients is complex and depends also on the heater length and the Rayleigh number. For a given heater length fraction, heat transfer coefficients decrease slightly with an increase in the aspect ratio. Also the number of convective cells at a given aspect ratio and heater length fraction is found to change with increasing Rayleigh number.

Chapter 5
Measurement of Heat Transfer Coefficients

Keywords Heater length ratio • Height ratio • Flux-based rayleigh number • Convective cells

In this chapter, measurements of steady-state Nusselt numbers in superposed fluid-porous layers with $\eta < 1$ and $\delta < 1$ are discussed and compared to computed values. These measurements point to the need for more exhaustive experimentation over an extended range of Rayleigh number, but they also provide entirely new data for that case where $\delta < 1$ and $\eta < 1$.

5.1 Design of the Experiment

A laboratory apparatus that is the thermal equivalent of the computational domain of Fig. 3.3 is shown in Fig. 5.1. The superposed porous and fluid sub-layers are bounded by the top and bottom walls with a fixed total height of $H = 3.81$ cm. The porous sub-layer comprises a level bed of 3 mm diameter spherical soda-lime glass beads ($k_s = 0.764$ W/m K, $\rho_s = 2500$ kg/m^3, $c_{p,s} = 918.2$ J/kg K). The relative height of the porous sub-layer is varied in several increments so that $0.5 \leq \eta \leq 1$. The minimum height of the porous sub-layer is six bead diameters to assure minimal wall effects.

Resistance heaters on the bottom of the porous sub-layer provide a constant heat flux boundary condition. Measurements of average heater temperature, T_H, are obtained with an array of thermocouples fixed to the heater surface. Overall heat losses are limited to one percent of input with a guard heater on the back of the bottom wall. The top of the apparatus is held at a constant temperature, T_C, with an attached heat exchanger. The top wall temperature is monitored by five thermocouples attached to its inner surface. In addition temperature measurements within the system are made at selected locations 1.9 cm above the base of the cell

A. Bagchi and F. A. Kulacki, *Natural Convection in Superposed Fluid-Porous Layers*, SpringerBriefs in Thermal Engineering and Applied Sciences, DOI: 10.1007/978-1-4614-6576-8_5, © The Author(s) 2014

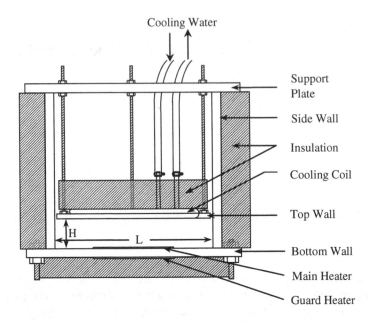

Fig. 5.1 Experimental apparatus. $L = 22.9$ cm, and $H = 3.81$ cm

by thirteen thermocouples attached to nylon filaments along the longitudinal and transverse centerlines. Additional details of the design and construction of the apparatus are given by Bagchi and Kulacki (2011).

In terms of the known heat flux and measurable quantities, the Nusselt and Rayleigh numbers are recast,

$$\text{Nu} = \frac{(q'' - q''_{\text{loss}})H}{k_{\text{eff}}\left(T_{H,\text{avg}} - T_{C,\text{avg}}\right)}, \tag{5.1}$$

$$\text{Ra} = \frac{g\beta H^3(T_{H,\text{avg}} - T_{C,\text{avg}})}{\nu_f \alpha_f}. \tag{5.2}$$

There are six primary independent parameters: η, δ, A, Da, λ, and Pr. In addition, the following parameters, which are assumed constant in the numerical simulations, must also be considered: ϕ, F, and σ. There are thus a total of nine independent controlling parameters for heat transfer. However, as seen in the numerical results, only the Rayleigh number, height ratio and heater length ratio significantly affect the overall heat transfer coefficient. As such, the effects of only these parameters are investigated (Table 5.1).

Four height ratios ($\eta = 0.5, 0.67, 0.75, 1$) and two heater length ratios ($\delta = 0.11, 0.44$) are addressed, and for each case $5 \times 10^6 < \text{Ra} < 10^8$. Because a single working height and solid–fluid combination is used, aspect ratio,

Table 5.1 Experimental parameters

	δ	η	H_1 (cm)	ΔT (°C)	Ra
Verification	0.44	1	3.81	3–40	5×10^6—10^8
Effect of η	0.44	0.5	1.91	3–40	5×10^6—10^8
		0.67	2.54		
		0.75	2.86		
Effect of δ	0.11	0.5	1.91	3–40	5×10^6—10^8
	0.44				

conductivity ratio and the Prandtl number remain constant (temperature variation of the thermal conductivity and Prandtl number are neglected). The porosity of the porous sub-layer varies from 0.36 to 0.39 and is therefore taken to be constant at ≈ 0.38, thus fixing the Darcy number at 4×10^{-6}. The Nusselt number is based on the effective thermal conductivity of the composite domain,

$$k_{\text{eff}} = k_f \left[(1 - \eta) + \frac{\eta}{\phi (1 - \phi) \lambda} \right]^{-1}. \tag{5.3}$$

It must be noted that the range of Rayleigh numbers used in the experiments does not overlap with the Rayleigh numbers used in the numerical calculations. The minimum Rayleigh number in the experiments could have been reduced further by either reducing the working height or reducing the temperature difference across the domain. However, the working height is fixed to ensure that the porous layer is at least six bead diameters thick at the lowest height ratio ($\eta = 0.5$). Similarly, the minimum temperature difference that is reliably measured is 3 °C. Therefore a few numerical simulations are performed up to Rayleigh numbers of 4×10^7 for comparison with experimental results. The comparison is discussed in detail in the next chapter.

Total heat loss is calculated by summing heat losses through the bottom and side walls using one-dimensional conduction models. Fluid properties are evaluated at the mean temperature across the total height of the system. The overall uncertainty in the Nusselt and Rayleigh numbers change with the Rayleigh number. Total uncertainties (~15 %) are higher at low Rayleigh number because of the small temperature difference between the heater surface and the top wall, as well a higher fraction of energy lost by conduction. For large Rayleigh numbers, uncertainties decrease to ~5–6 % in Ra and ~3–4 % in Nu. An analysis of experimental uncertainty is given by Bagchi (2010).

To validate the present experimental setup, heat transfer coefficients are measured for a fully porous layer ($\eta = 1$) and $\delta = 0.44$ and 0.11. Results of the experiments are shown in Fig. 5.2 with the numerical and experimental results of Rajen and Kulacki (1987), Kulacki and Rajen (1991) and Lai and Kulacki (1991). Specific parameters are listed in Table 5.2. Above the critical point, agreement with the numerical simulation results of Rajen and Kulacki (1987) is very good over the entire Rayleigh number range with a maximum difference of ~4.5 percent at $Ra_m^* = 620$. The present results also agree well with the experimental

Fig. 5.2 Comparison of
current experimental data
with the numerical and
experimental results of Rajen
and Kulacki (1987) and
Kulacki and Rajen (1991)

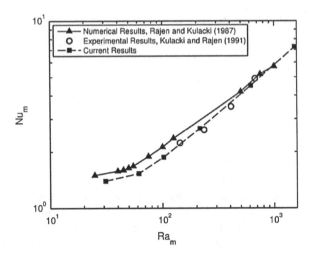

Table 5.2 Parameters
used for studies on
validation

	Present experiments	Rajen and Kulacki (1987)	Lai and Kulacki (1991)
Height ratio, η	0	0	0
Heater length ratio, δ	0.44, 0.11	0.5	0.14
Aspect ratio, A	6	6	21
Darcy number, Da	4.04×10^{-6}	2×10^{-6}	3.4×10^{-6}

results of Kulacki and Rajen (1991). This latter difference is to be expected
considering that the heater length ratio in the present experiments ($\delta = 0.44$) is
slightly less than that used by Rajen and Kulacki ($\delta = 0.5$). Figure 5.3 shows the
comparison of the present results for $\delta = 0.11$ and it can be seen that the agreement is very good over the entire range of Rayleigh numbers. It must be noted
that the heater length ratio in the present study is slightly smaller than that used
by Lai and Kulacki (1991). The results also agree well with the results of their
numerical simulations.

5.2 Measured Nusselt Numbers

Heat transfer results for the present problem are shown in Figs. 5.4 and 5.5.
Figure 5.4 shows Nusselt-Rayleigh number data with regression fits for $\delta = 0.44$
and $\eta = 0.5$, 0.67, 0.75 and 1. The heat transfer data is adequately correlated by
a power-law relation of the form $Nu = C \times Ra^n$, where C and n are constants.
Table 5.3 summarizes values of C and n for several geometric parameters.

Fig. 5.3 Comparison of current experimental results with the numerical and experimental results of Lai and Kulacki (1987). $\eta = 1$

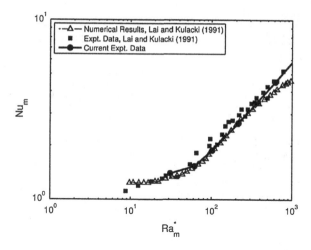

Fig. 5.4 Nusselt-versus-Rayleigh number. $\delta = 0.44$

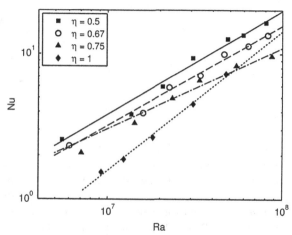

Fig. 5.5 Nusselt-versus-Rayleigh number. $\eta = 0.5$

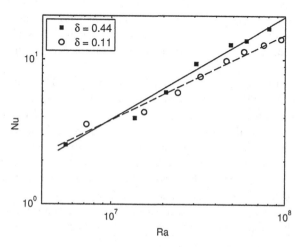

Table 5.3 Curve fit parameters

Parameters	Curve-fit equation	R^2
$\delta = 0.44$, $\eta = 0.5$	$Nu = (3.964 \times 10^{-5}) \times Ra^{0.7121}$	0.9823
$\delta = 0.44$, $\eta = 0.67$	$Nu = (5.017 \times 10^{-5}) \times Ra^{0.6866}$	0.9891
$\delta = 0.44$, $\eta = 0.75$	$Nu = (3.934 \times 10^{-4}) \times Ra^{0.5557}$	0.9727
$\delta = 0.11$, $\eta = 0.5$	$Nu = (2.831 \times 10^{-4}) \times Ra^{0.59}$	0.9749

Fig. 5.6 Non-dimensional horizontal temperature profiles cell at the horizontal mid-plane. The position of the heater is also shown on the x-axis

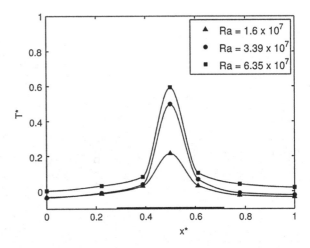

Figure 5.6 also shows that the overall Nusselt number increases with a decrease in η at a given Rayleigh number. This result is attributed to the tendency toward more vigorous fluid motion in the fluid sub-layer with its increasing height and is also in qualitative agreement with the numerical simulation and prior numerical and experimental results for natural convection in uniformly heated fluid-superposed porous layers (Poulikakos 1986; Chen and Chen 1992; Prasad and Tian 1990; Prasad et al. 1991). There is some overlap in the Nusselt number data near at the lower Rayeligh numbers and is most likely due to larger uncertainty in the Nusselt number (~15 %). Given the observed trends in Nusselt number with η and δ, a reasonable conjecture is that the effect of η is decoupled from the horizontal extent of the heat source.

Figure 5.5 shows the Nusselt-versus-Rayleigh number data for $\eta = 0.5$ and $\delta = 0.11$ and 0.44 with regression fits. At low Rayleigh number, Nusselt numbers for $\delta = 0.11$ are larger than those for $\delta = 0.44$. At higher Rayleigh numbers however, the trend is reversed and the Nusselt numbers for $\delta = 0.44$ are larger. Overall, however, the data sets are fairly close to each other indicating that a change in the size of the heater does not have a significant effect on the heat transfer coefficients. This observation is in qualitative disagreement with the results of the numerical simulation results which show that the Nusselt number over the heater surface increases with a decrease in the size of the heater. The increase in the Nusselt number with decreasing heater size has been observed in prior studies of natural convection in both porous and fluid layers heated locally from below (Rajen and Kulacki 1987; Lai and Kulacki 1991; Papanicolau and Gopalakrishna 1995). However a clear trend has only been seen in numerical studies while observable differences are much less clear in experimental data, particularly at high Rayleigh number. Because the current data is well within the super-critical regime, a lack of noticeable increase in the Nusselt number with decrease in the heater length ratio may not represent a direct contradiction of the numerical predictions, especially as the simulations explore a much lower Rayleigh number regime. Thus further experiments are essential to clarify the effect of the heater length.

5.3 Temperature Profiles

Temperature profiles are measured along the horizontal mid-plane above the heater surface along the heater length and in the transverse direction. Figure 5.6 shows the transverse direction dimensionless temperature profiles for $A = 6$, $\eta = 0.67$ and $\delta = 0.44$ at three Rayleigh numbers. The profiles show a distinct plume like flow at the centerline that is more pronounced as Rayleigh number is increased. A single plume is present, which is in agreement with the numerical simulation. The most important aspect of the temperature profile, however, is the vertical location of the thermal plume within the composite domain. Because the thermocouples are located at the horizontal mid-plane of the apparatus and $\eta = 0.67$, they are embedded within the porous sub-layer below the interface. The presence of a plume like flow inside the porous sub-layer is evidence that convective motion is present and confirms the presence of penetrative convection. It also shows that the flow inside the porous sub-layer is not conduction dominated.

Fig. 5.7 Non-dimensional temperature profiles along the depth of the convection cell at the horizontal mid-plane

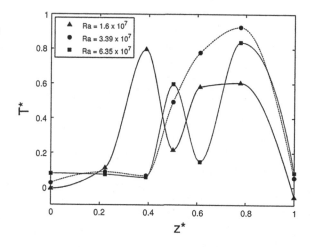

Longitudinal temperature profiles shown exhibit a complex flow pattern (Fig. 5.7). Interestingly, the temperature field changes with increasing Rayleigh number. For Ra = 1.6×10^7, a single well developed thermal plume and another partially developed thermal plume are seen. At Ra = 3.39×10^7, only a single plume can be seen. Further increase in Ra creates two well developed thermal plumes. It also reveals a flow structure that is highly asymmetric and suggests three-dimensionality. These observations are in agreement with the experiments of Chen and Chen (1989).

Chapter 6
Discussion

Keywords Numerical-experimental   comparison • Two-domain   formulation •
One-domain formulation • Numerical issues • Current research state

In this chapter numerical and experimental results are compared in order to validate the numerical model and its solution. The discrepancies between the numerical and experimental results are analyzed to understand the underlying cause of the
disagreement. In particular, the validity of the one-domain formulation is discussed.
Numerical results available in the literature are compared, and the discrepancies
among different modeling approaches are reviewed. Thereafter the importance of
the present study in the context of the overall problem of convection in composite
fluid-porous systems is discussed.

6.1 Comparison of Numerical and Experimental Results

To accurately compare the results of the numerical and the experimental studies,
numerical solutions have been obtained with independent controlling parameters of
the experiments. These are: $A = 6$, $Da = 4.04 \times 10^{-6}$, $\phi = 0.36$, $\delta = 0.44$, $\lambda = 1$,
$F = 0.52$, $\sigma = 0.5$, $\eta = 0.5, 0.75$. Figures 6.1 and 6.2, and Table 6.1, show a comparison of the numerical and experimental results. For both $\eta = 0.5$ and $\eta = 0.75$,
the difference between the numerical predictions and experimental measurements
is significant in the Rayleigh number range where the two data sets overlap. This
mismatch is more pronounced for $\eta = 0.5$ than for $\eta = 0.75$, but in both cases the
percentage difference between the numerical and experimental data is $\geq 60\ \%$. It
can also be seen that the slope of the experimental data fit line is much steeper than
that of the numerical data. The fundamentally different nature of the two curves
indicates that there is a fundamental underlying discrepancy between the numerical and experimental results. This discrepancy is also seen for the other two sets of
experiments.

A. Bagchi and F. A. Kulacki, *Natural Convection in Superposed Fluid-Porous Layers,*   67
SpringerBriefs in Thermal Engineering and Applied Sciences,
DOI: 10.1007/978-1-4614-6576-8_6, © The Author(s) 2014

Fig. 6.1 Comparison of
numerical and experimental
data for $\delta = 0.44$, $\eta = 0.5$

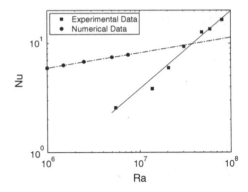

Fig. 6.2 Comparison of
numerical and experimental
data for $\delta = 0.44$, $\eta = 0.75$

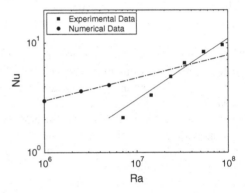

Table 6.1 Comparison of
experimental and numerical
Nusselt numbers

η	Ra	Nu (Experimental)	Nu (Simulation)	% Error
0.5	5.47	2.219	7.52	70.5
	7.5	2.79	7.86	64.5
0.75	7.45	1.789	4.45	59.82

The nature of the discrepancy between the numerical and experimental results
indicates that the mismatch cannot be simply attributed to factors such as the
uncertainty in the measurements and the grid convergence of the numerical solution. Simulations with finer grids and a more stringent convergence criterion produce no significant change in the results. Further, because both the computer code
and the experimental setup are verified by comparison with literature results for
well-established test cases (Rayleigh-Bénard convection and the Horton-Rogers-
Lapwood problem), the observed difference cannot be attributed errors in post-
processing of the data. It can therefore be concluded that the observed difference
points to a discrepancy that is fundamental in nature. To investigate the cause

behind the observed difference, attention is focused on the problem definition, the experimental setup, and the mathematical formulation.

The first reason for the observed discrepancy between the numerical predictions and experimental results may be due to the differences in the boundary conditions for the two studies. In particular, attention needs to be focused on the boundary conditions along the lower surface. The problem investigated via numerical simulation is that of two-dimensional natural convection in a fluid-superposed porous layer with a localized isothermal heat source at the base. The experiments on the other hand are conducted in a three-dimensional enclosure with a strip heater supplying a constant heat flux to the porous layer. The boundary conditions at the heater surface are therefore different for the two problems. The difference in boundary conditions can be better understood by comparing the temperature distributions at the heater and top surfaces for a single experiment ($\delta = 0.44$, $\eta = 0.5$; Table 6.2). It can be seen that while a nearly isothermal condition exists at the top plate, the same is not true for the heater surface. For most runs, the percentage standard deviation (SD in Table 6.1) of the heater surface temperature is >10 %. This variation in temperature across the heater, which is in part responsible for the high uncertainties in Nu and Ra is one source of the mismatch. Another point of discrepancy lies in the boundary conditions at the base outside the heater area. The formulation of the numerical solution assumes that base area outside the heated region is adiabatic. This condition is, however not replicated in the apparatus. A numerical solution of the conduction problem in the base plate shows that there is a small heat flow from the area outside the strip heater into the porous layer. Although this heat flow is a very small fraction of the total energy supplied by the heaters, it nevertheless causes a part of the area outside the strip heater to be non-adiabatic.

These differences in the boundary conditions indicate that a difference between the numerical and experimental Nusselt numbers can be expected. However, it is worthwhile investigating whether all of the observed discrepancy occurs solely because of this reason. As there are no published studies on convection in fluid superposed porous layers with a constant heat flux boundary condition at the base, a direct estimation of the effect of the boundary conditions on the Nusselt and Rayleigh numbers is not possible. There are, however, several studies of natural convection in porous layers with localized isothermal and constant flux heat sources at the base. Because the porous layer convection problem is a special case of the

Table 6.2 Temperatures on the heater and upper surfaces with standard deviation (SD). $\eta = 0.75$, $\delta = 0.44$

$T_{H,avg}$	SD (%)	$T_{C,avg}$	SD (%)
29.57	4.28	23.53	0.24
33.98	9.99	22.83	0.38
39.11	13.08	22.94	0.61
43.84	13.66	23.23	0.90
52.28	13.37	25.04	1.08
61.79	14.3	24.65	1.61

present problem for $\eta = 1$, results from these studies can be used to understand the effect of the boundary conditions on the overall heat transfer coefficients.

A comparison between the numerical results of Prasad and Kulacki (1987) with the numerical and experimental results of Rajen and Kulacki (1987) and Kulacki and Rajen (1991) is chosen to investigate the effects of the boundary conditions. The simulations of Prasad and Kulacki are performed with $A = 10$, $\delta = 0.5$, and a centrally positioned isothermal heater. The numerical and experimental results of Rajen and Kulacki use $A = 9.6$, $\delta = 0.5$, and a centrally positioned constant flux heater. Except for the boundary condition at the heater surface, the two studies have nearly identical problem formulations and boundary conditions and are ideally suited for comparison. It must be noted that Prasad and Kulacki define the Rayleigh number, Ra_m, based on the temperature difference across the heated and cooled surfaces, and Rajen and Kulacki use a flux-based definition for the Rayleigh number Ra_m^*. The two Rayleigh numbers are related, $Ra_m^* = Ra \times Nu_m$, where Nu_m is based on the stagnant conductivity of the porous sub-layer. Thus the Nu_m-versus-Ra_m^* data of Rajen and Kulacki (1987) and Kulacki and Rajen (1991) are converted to Nu_m-versus-Ra_m data for comparison. This step is similar to the experimental data reduction procedure adopted in the experiments where the temperature-based Rayleigh number is used for the constant flux heater.

Comparison of the data for the above studies is shown in Fig. 6.3. The differences in the data sets are readily apparent, confirming that heat transfer coefficients change with a change in the boundary conditions. Although the data sets are different, the Nu_m-versus-Ra_m curves have almost identical overall character although the slope of the curve for the constant heat flux data is higher. In addition, for a given Ra_m, the value of Nu_m is higher for the constant heat flux boundary condition data. If these observations are extrapolated to the present problem, one would expect to see identical characteristics in the numerical and experimental Nu-versus-Ra curves and experimental data over predicting Nu over the entire range of Ra. It can be immediately seen that none of these characteristics is observed. Thus, it can be concluded the observed discrepancy between the

Fig. 6.3 Numerical and experimental results of Prasad and Kulacki (1987) and Rajen and Kulacki (1987)

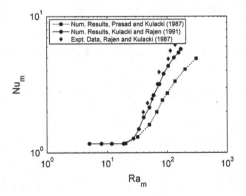

numerical and experimental results of this study cannot be attributed to the differences in the boundary conditions of the two studies.

The other possible source of discrepancy between the simulation and experimental data may be the mathematical formulation of the problem for the numerical solution. In particular, attention must be focused on the one-domain formulation that is used here. It may be well that the limitations of this particular formulation are responsible for the observed lack of agreement with the experimental results. To further explore this issue, it is important to first understand the validity and accuracy of the one-domain formulation. An extensive review of the literature reveals that there has been no thorough experimental validation of the one-domain formulation for the present problem. The only reported comparison of simulations using this formulation to experimental data has been reported by Kim and Choi (1996), who determine the critical wave number for different height ratios and find good agreement with the experimental data of Chen and Chen (1992).

They also determine the Nusselt numbers for Rayleigh number up to two times the critical value at $\eta = 0.91$ and find good agreement with the Chen and Chen results. However, they do not validate their numerical solution at Rayleigh number much higher than the critical point. Thus, it is quite possible that the model may have some deficiencies, especially in the high Rayleigh number range where its validity is yet to be demonstrated. To explore this possibility, simulations are performed with the present code for the case of convection with a uniformly heated base and the results are compared with the experimental results of Prasad and Tian (1990). The parameters are $A = 2$, $Da = 3.71 \times 10^{-6}$, $\phi = 0.396$, $\lambda = 6.8$, $Pr = 8835$, $F = 0.56$, $\sigma = 0.7$, $\eta = 0.5, 0.8$, and are suitable for a randomly packed layer of 6 mm diameter glass beads saturated with Dow Corning 200 Silicone oil.

Comparison of the results of the simulations with the data of Prasad and Tian (1990) is shown in Fig. 6.4 and Table 6.3. The marked discrepancy between the two sets of data is clearly apparent. Although Prasad and Tian do not provide information on the uncertainties in their measurements, it can be safely said that

Fig. 6.4 Comparison of simulations results for $\delta = 1$ and $\eta = 0.5, 0.8$ with the experimental results of Prasad and Tian (1990)

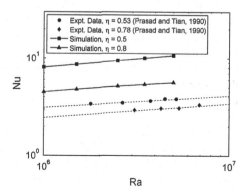

Table 6.3 Comparison of numerical simulations with data of Prasad and Tian (1990)

Ra	$\eta = 0.53$			$\eta = 0.78$		
	Nu (Sim.)	Nu (Expt.)	% Diff.	Nu (Sim.)	Nu (Expt.)	% Diff.
1×10^6	8.08	3.11	61.42	4.51	2.45	45.60
1.5×10^6	8.67	3.27	62.25	4.81	2.60	45.95
2.5×10^6	9.48	3.48	63.21	5.17	2.80	45.86
3.5×10^6	10.04	3.63	63.81	5.41	2.94	45.62
5×10^6	10.66	3.79	64.42	5.64	3.09	45.06

such a large disagreement cannot be simply due to the uncertainties. In Table 6.3 it can be seen that the difference between the simulation and experimental data is larger for $\eta = 0.5$ than for $\eta = 0.8$. This is identical to what is observed for the present experimental measurements. It can therefore be said that the discrepancy between the theoretical and experimental results is higher at lower height ratios. The comparison with the data of Prasad and Tian (1990) indicates that the present implementation of the one-domain model gives physically unrealistic data in the high Rayleigh number range.

To get a better idea of the influence of the effect of the modeling on the heat transfer results, results from all numerical studies available in the literature are compared with the present simulation for $\delta = 1$ in Fig. 6.5. The parameters used in each of the compared studies are listed in Table 6.4, and it can be seen that the simulation parameters used in all the studies are more or less identical. The only parameters that differ somewhat among all studies are the aspect ratio, A, and the Darcy number, Da. However, as seen in Chap. 4, these parameters do not significantly affect the overall heat transfer coefficients. As such, the comparison can give a good idea of the effect of the modeling approach. It must be noted that three modeling approaches are compared here: (i) the two-domain approach which uses Darcy's law in the porous layer and the Beavers-Joseph slip boundary condition at the interface, (ii) a two- domain formulation which uses Brinkman's extension to Darcy's law in the porous layer and the continuity of shear stress and tangential velocity at the interface (Neale and Nader 1974), and (iii) a one-domain formulation which does not involve any explicit specification of the interfacial boundary conditions (Poulikakos 1986).

Figure 6.5 shows how a given modeling approach drastically affects the overall Nusselt-versus-Rayleigh number relation. In particular, it can be seen how different results are obtained using the one-domain formulation. While, the present results show excellent agreement with the results of Poulikakos (1986), they differ significantly from those of Kim and Choi (1996) although all three studies use the one-domain formulation. This indicates that the limitations of the one-domain model may lie not in the mathematical formulation of the model but rather in its numerical implementation.

The most conclusive evidence for this hypothesis can be found the publications of Hirata and co-workers (2006, 2007a, b, 2009). This fact was pointed out earlier in Chap. 2, but its significance can be best understood in the present context. To briefly recall the earlier discussion. Hirata et al. (2006) first published a

Fig. 6.5 Comparison of present results for $\delta = 1$ with all published numerical studies

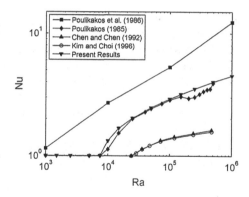

Table 6.4 Parameters used in the different numerical studies

Parameter	Poulikakos et al. (1986)	Poulikakos (1985)	Chen and Chen (1992)	Kim and Choi (1996)	Present study
Da	10^{-6}	10^{-4}	2.2×10^{-6}	2.2×10^{-6}	10^{-6}
η	0.5	0.5	0.5	0.5	0.5
Pr	7	7	6.26	6.26	7
λ	1	1	1.62	1.62	1

study in which they investigated the effect of the mathematical formulation on the prediction of the stability criterion for the onset of convection. They compare the three different modeling approaches mentioned above and show that the modeling approach adopted has a profound effect on the prediction of the stability criterion. They find that while both the two-domain formulations give almost identical marginal stability curves, the curves obtained using the one-domain formulation are markedly different. The stability curves for the one-domain formulation do not exhibit the signature bi-modal character that is unique to the problem of convection in fluid-superposed porous layers. Also the critical Rayleigh numbers predicted by the one-domain formulation differ from those predicted by the two-domain approaches by 30–40 % at different height ratios. This is contrast to the findings of Zhao and Chen (2001) who investigate the stability problem using the one-domain formulation and find that the stability curves show the expected bi-modal character. These findings further point to the fact that the results obtained using the one-domain formulation strongly depend on the particular solution technique adopted.

To further investigate the source of this discrepancy, Hirata et al. (2009) re-examine the same problem with a different approach to the numerical treatment of the interface conditions. They incorporate the hypothesis proposed by Kataoka (1986) that the average properties of the porous medium, such as the porosity, permeability, and effective diffusivity are Heaviside step functions and hence their differentiation must be considered in the meaning of distributions. Using this approach they find that marginal stability curves for the one domain and two domain approaches are almost identical and show almost the same bimodal behavior. Based on their results, Hirata et al. conclude that the one- and two-domain approaches are identical provided that the one-domain approach is properly interpreted mathematically, i.e., in the meaning of distributions. Based on this evidence it can be concluded that the limitation of the one-domain formulation lies in its implementation in a numerical code. This conclusion is further supported by the fact that mathematical formulation of the one-domain model is identical to that of the two-domain model which assumes continuity of velocity, normal stress, shear, heat flux, and temperature at interface. The model is different only in the sense that it does not involve the explicit specification of the interface boundary conditions.

It can therefore be concluded that special care must be taken while numerically implementing the one-domain formulation. In particular, the discretization equations for the interfacial control volumes must be treated separately so that the differentiations of the Heaviside step functions are correctly performed. It must be mentioned here that the approach adopted by Hirata et al. (2009) must not be considered as an integral part of the one-domain model. Other reported results obtained using the one-domain model obtain good agreement with experimental results without using specific mathematical constructs (Kim and Choi 1996). Also, their approach has not been verified by any subsequent study. Of special note is that their study is restricted to the determination of the stability criterion. The extension of this approach to the high Rayleigh number domain of the problem remains for investigation. Most importantly though, all existing studies devoted to the stability problem rely on a single set of validating experiments (Chen and Chen 1992). These experiments have not been replicated. Therefore, it cannot be conclusively said that the mathematical approach proposed by Hirata et al. (2009) can resolve the conundrum regarding the most accurate mathematical model for the fluid-porous layer interface.

It must be pointed out that the discrepancies observed between the results of the one-and two-domain models seem to be restricted to the particular problem of natural convection in horizontal composite fluid-porous domains heated from below. In other studies on natural convection heat transfer in composite fluid-porous domains, the validity of both formulations has been demonstrated appropriately. For example Beckerman and co-workers (Beckerman et al. 1987, 1988) used the one-domain formulation to investigate natural convection in horizontal and vertical composite domains heated on the sides and obtained excellent agreement with experimental measurements. Similarly, Singh and Thorpe (1995) studied natural convection in horizontal fluid-superposed porous layers heated on the sides using the two-domain (with both the Beavers-Joseph and continuity conditions) and the one-domain formulations and found that all three modeling approaches gave nearly identical results. Thus, it appears that the issues associated with the implementation of the one-domain formulation are not inherent in the method itself but its application to the present problem.

A question that must be answered at this point is why the one-domain model is chosen, considering the confusion that exists regarding its implementation. The reasoning behind using the model is two-fold. First, the model is much simpler to use and implement than other modeling approaches as a single set of governing equations can be used for the entire domain and discretization equations for the boundary conditions do not have to be separately derived. Second, from a physical perspective, the one-domain concept is a generic model. It may be recalled that the crux of the model is that the porosity of the porous layer acts as a switching parameter that selects the applicable form of the governing equations based on whether the solution domain lies in the fluid or porous layers. Here, the porosity is a binary parameter and can only assume the values zero and unity, but there is no restriction mathematically on the values that it can take. This allows problems involving nonhomogenous, fissured, and layered porous media to be handled easily by simply assigning spatial or functional values to the porosity. This analytic property makes the one-domain formulation very powerful. When coupled with its ease of programming and implementation, it is easily seen why this technique is so attractive and why so much research has been devoted to reconcile this formulation with existing two-domain models. However, before the one-domain formulation can be universally applied its validity must be clearly demonstrated by showing its equivalence to the two-domain formulations and comparison to experimental measurements. The authors are currently involved in developing a mathematically consistent and physically realistic implementation of the one-domain formulation that can be validated by experimental measurements.

6.2 Summary

Our objective has been to develop a more complete picture of convection in superposed fluid and porous layers. The combination of numerical and experimental methods is apparently the first attempt at validation of the one-domain formulation of the governing equations. Such a validation of theoretical studies is generally absent in the literature owing to the difficulty in designing experiments which can measure heat transfer rates at Rayleigh numbers accessible to simulation. Further our numerical simulations have highlighted some of the challenges that arise when implementing the one-domain formulation.

The contributions of this study can be better understood by considering the results in the overall context of the problem of convection in fluid-superposed porous layers. As can be gathered from the discussion in the preceding section, the vast differences in the predictions of different numerical studies highlight some of the challenges that exist in accurately modeling the superposed system. Although the first numerical solution was published over 25 years ago, no accepted set of results is yet available even for the case of a uniformly heated base. This is especially true for the high Rayleigh number regime where lack of any experimental validation has made it impossible to verify the accuracy of any of the proposed modeling approaches.

Similarly experimental studies have focused on studying very small sections of the entire convection heat transfer regime (Table 6.5). This is best illustrated in

Table 6.5 Parameters of the existing experimental studies

Study	A	Da	Pr	λ
Prasad et al. (1990)	2	3.27×10^{-6}	8835	6.81
Prasad et al. (1991)	2	1.3×10^{-5}	578	1.19
Prasad (1993)	2	8.2×10^{-5}	8835	1
Chen and Chen (1992)	6	2.2×10^{-6}	6.26	1.62
Steven (2006)	0.5	3.1×10^{-6}	6.26	1
Present study	6	4.4×10^{-6}	6.26	1

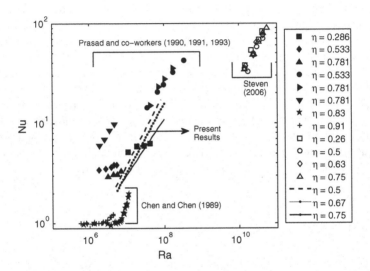

Fig. 6.6 Comparison of existing experimental studies

Fig. 6.6 which shows a compilation of existing experimental data. It can be readily seen that the data are scattered over a wide range of Rayleigh numbers and do not conform to any well-defined pattern. This is in contrast to the Rayleigh-Bénard and Horton-Rogers-Lapwood problems for which the experimental data follow a coherent pattern which has suggested a fundamental underlying relation between Nusselt and Rayleigh numbers. Part of this can be attributed to the complexity of the superposed problem because of the large number of independent controlling parameters. However, part of the reason why the overall nature of the problem is not yet well understood is that previous studies have focused on studying very specific aspects of it. Taken in this context, the combined numerical and experimental approach reported in this monograph represents an important step toward fundamental understanding over a large portion of the entire heat transfer regime.

References

Aichlmayr HT, Kulacki FA (2006) The effective thermal conductivity of saturated porous media. In: green G et al (eds) Advances in heat transfer, vol 39 Academic Press, New York, pp 377–460

Allen MB (1984) Collocation techniques for modeling compositional flows in oil reservoirs. Springer, New York

Allen MB, Khosravani A (1992) Solute transport via alternating-direction collocation using the modified method of characteristics. Adv Water Resour 15:125–132

Bagchi A (2010) Natural convection in horizontal fluid-superposed porous layers heated locally from below. University of Minnesota, Doctoral Dissertation

Bagchi A, Kulacki FA (2011) Natural convection in fluid superposed porous layers with localized heat sources. Int J Heat Mass Trans 54:3673–3682

Bagchi A, Kulacki FA (2012) Experimental study of natural convection in fluid-superposed porous layers heat locally from below. Int J Heat Mass Trans 55:1149–1153

Beavers GS, Joseph DD (1967) Boundary conditions at a naturally permeable wall. J Fluid Mech 30:197–207

Beckermann C, Ramadhyani S, Viskanta R (1987) Natural convection flow and heat transfer between a fluid and a porous layer inside a rectangular enclosure. ASME J. Heat Trans 109:363–370

Beckermann C, Viskanta R, Ramadhyani S (1988) Natural convection in vertical enclosures containing simultaneously fluid and porous layers. J Fluid Mech 186:257–284

Buretta JP, Berman AS (1976) Convective heat transfer in a liquid saturated porous layer. ASME J Appl Mech 98:249–254

Caltagirone JP (1975) Thermoconvective instabilities in a horizontal porous layer. J Fluid Mech 72:269–287

Carr M, Straughan B (2003) Penetrative convection in a fluid overlying a porous medium. Adv Water Resour 26:262–276

Chen F (1990) On the stability of salt-finger convection in superposed fluid and porous layers. ASME J Heat Trans 112:1088–1092

Chen F (1991) Throughflow effect on convective instability in superposed fluid and porous layers. J Fluid Mech 23:113–133

Chen F, Chen CF (1988) Onset of finger convection in a horizontal porous layer underlying a fluid layers. ASME J Heat Trans 110:403–409

Chen F, Chen CF (1989) Experimental investigation of convective stability in a superposed fluid and porous layer when heat from below. J Fluid Mech 207:311–321

Chen F, Chen CF (1992) Natural convection in superposed fluid ad porous layers. J Fluid Mech 234:97–119

A. Bagchi and F. A. Kulacki, *Natural Convection in Superposed Fluid-Porous Layers*, 77
SpringerBriefs in Thermal Engineering and Applied Sciences,
DOI: 10.1007/978-1-4614-6576-8, © The Author(s) 2014

Chen F, Chen CF, Pearlstein AJ (1991) Convective instability in superposed fluid and anisotropic porous layers. Phys Fluids A: Fluid Dyn 3:556–565

Chen F, Hsu LH (1991) Onset of thermal convection in an anisotropic and inhomogeneous porous layer underlying a fluid layer. J. Appl Phys 69:6289–6301

Chen F, Lu JW (1992) Variable viscosity effects on convective instability in superposed fluid and porous layers. Phys Fluids A: Fluid Dyn 4:1936–1944

Chen QS, Prasad V, Chatterjee A (1999) Modeling of fluid flow and heat transfer in a hydrothermal crystal growth system: Use of fluid-superposed porous layer theory. ASME J Heat Trans 121:1049–1058

Cheng P (1978) Heat transfer in geothermal systems. In: Irvine T, Hartnett JP (eds) Advances in heat transfer, vol 14. Academic Press, New York, pp 1-105

Clever RM, Busse FH (1974) Transition to time-dependent convection. J Fluid Mech 6:625–635

Combarnous MS, Bories A (1975) Hydrothermal convection in saturated porous media. Adv Hydrosci 10:231–307

Curran MC, Allen MB (1990) Parallel computing for solute transport models via alternating direction collocation. Adv Water Resour 13:70–75

Eklund H (1963) Freshwater: temperature of maximum density calculated from compressibility. Sci 142:1457–58

Elder JM (1967) Steady free convection in a porous medium heated from below. J. Fluid Mech 27:29–48

Ewing RC (1996) Multidisciplinary interactions in energy and environmental engineering. J Comput Appl Math 74:193–215

Gobin D, Goyeau B (2008) Natural convection in partially porous media: a brief overview. Int J Num Meth Heat Fluid Flow 18:465–490

Hayase T, Humphrey JAC, Greif R (1992) A consistently formulated QICK scheme for fast and stable convergence using finite-volume iterative calculation procedures. J Comp Phys 98:108–118

Hirata SC, Goyeau B, Gobin D, Cotta RM (2006) Stability of natural convection in superposed fluid and porous layers using integral transforms. Num Heat Trans 50:409–424

Hirata SC, Goyeau B, Gobin D (2007) Stability of natural convection in superposed fluid and porous layers: influence of the interfacial jump boundary condition. Phys fluids 19:058102

Hirata SD, Goyeau B, Gobin D, Carr M, Cotta (2007) Linear stability of natural convection in superposed fluid and porous layers: influence of interfacial modeling. Int J Heat Mass Trans 50:1356–1367

Hirata SC, Goyeau B, Gobin D, Chandesris M, Jamet D (2009) Stability of natural convection in superposed fluid and porous layers: equivalence of the one- and two-domain approaches. Int. J Heat Mass Trans 52:533–536

Hollands KGT, Raithby GD, Konicek L (1975) Correlations equations for free convection heat transfer in horizontal layers of air and water. Int J Heat Mass Trans 18:879–884

Horton CW, Rogers FT (1945) Convection current in a porous medium. J Appl Phys 16:367–370

In: Heat transfer in geophysical and geothermal systems, Vafai K et al (eds) American Society of Mechanical Engineers, HTD-Vol. 76:27-36

Jendoubi S, Kulacki FA (1999) Convection in layered porous media In: A comparison of boundary heating methods. Proc ASME/JSME Joint Therm Eng Conf, Paper No. AJTE99-6281, American Society of Mechanical Engineers, New York

Jones IP (1973) Low Reynolds number flow past a porous spherical shell. Proc. Camb Phil Soc 73:231–238

Kataoka I (1986) Local instant formulation of two phase flow. Int J. Multiph Flow 12:745–758

Kato Y, Masuoka Y (1967) Criterion for the onset of convective flow in a fluid in a porous medium. Int J Heat Mass Trans 10:297309

Kaviany M (1991) Principles of heat transfer in porous media. Springer, New York

Kazmierczak M, Muley A (1994) Steady and transient natural convection experiments in a horizontal porous layers: effect of a thin top fluid layers and oscillatory bottom wall temperature. Int J Heat Mass Trans 15:30–41

Kim SJ, Choi CY (1996) Convective heat transfer in porous an overlying fluid layers heated from below. Int J Heat Mass Trans 39:319–329

Kulacki FA, Rajen G (1991) Buoyancy induced flow and heat transfer in fissured media. In: Kakac S et al (eds) Convective heat transfer in Porous Media. Ser E: Appl Sci, Kluwer, Dodrecht 109:465–498

Kunii D, Smith JM (1960) Heat transfer characteristics of porous rocks. AIChE J 6:71–78

Lai FC, Kulacki FA (1987) Natural convection in layered porous media partially het from below

Lai FC, Kulacki FA (1991) Experimental study of free and mixed convection in horizontal porous layers locally heated from below. Int. J Heat Mass Trans 34:535–541

Lapwood ER (1948) Convection of a fluid in a porous medium. Proc Camb Phil Soc 44:508–521

Leonard BP (1979) A stable and accurate convective modeling procedure based on quadratic upstream interpolation. Comput Meth Appl Mech Eng 19:59–98

Matthews PC (1988) A model for the onset of penetrative convection. J Fluid Mech 188:571–583

Neale G, Nader W (1974) Practical significance of brinkman extension of Darcy's law: coupled parallel flow within a channel and boundary porous medium. Can J Chem Eng 52:472–478

Nield DA (1977) Onset of convection in a fluid layer overlying a layer of a porous medium. J Fluid Mech 81:513–522

Nield DA, Bejan A (2006) Convection in porous media, 3rd edn. Springer-Verlag, New York

Ochoa-Tapia JA, Whitaker S (1995) Momentum transfer at the boundary between a porous medium and a homogeneous fluid–II comparison with experiments. Int J Heat Mass Trans 38:2647–2655

Ochoa-Tapia JA, Whitaker S (1995) Momentum transfer at the boundary between a porous medium and a homogeneous fluid–I theoretical development. Int J Heat Mass Trans 38:2635–2646

Papanicolau E, Gopalakrishna S (1995) Natural convection in shallow, horizontal air layers encountered in electronic cooling. J. Electron Packag 17:307–316

Patankar S (1980) Numerical heat transfer and fluid flow. Hemisphere, Washington

Platten JK, Legros JC (1984) Convection in liquids. Springer-Verlag, Berlin

Poulikakos D (1986) Buoyancy driven convection in a horizontal fluid layer extending over a porous substrate. Phys Fluids 29:3949–3957

Poulikakos D, Bejan A, Selimos B, Blake KR (1986) High Rayleigh number convection in a fluid overlying a porous bed. Int J Heat Fluid Flow 7:109–116

Prasad V (1991) Convective flow interaction and heat transfer between fluid and porous layers. In: Kakac S, Kulacki FA, Viskanta R (eds), proceedings of NATO ASI, convective heat and mass transfer in porous media, Kluwer, Boston, 563–615

Prasad V (1993) Flow instabilities and heat transfer in fluid overlying horizontal porous layers. Exp Therm Fluid Sci 6:135–146

Prasad V, Brown K, Tian Q (1991) Flow visualization and heat transfer experiments in fluid-superposed packed beds heat from below. Exp Therm Fluid Sci 4:12–24

Prasad V, Kulacki FA (1986) Effects of the size of heat source on natural convection in horizontal porous layers heat from below. In: Tien CL et al (eds) Heat transfer vol 5, Proc 8th Int Heat Trans Conf. Hemisphere, Washington, pp 2677-2682, 1986

Prasad V, Kulacki FA (1987) Natural convection in horizontal porous layers with localized heating from below. ASME J Heat Trans 109:795–798

Prasad V, Kulacki FA, Keyhani M (1985) Natural convection in porous media. J Fluid Mech 150:89–119

Prasad V, Tian Q (1990) An experimental study of thermal convection in fluid-superposed porous layers heat from below. In: Hetsroni et al (eds) Heat Transfer 1990, Proc 9th Int Heat Trans Conf, Hemisphere, New York, 5:207-212

Quertatani N, Cheikh NB, Beya BB, Lili T (2008) Numerical simulation of two-dimensional Rayleigh-Bénard convection in an enclosure. Comp Rend Mech 336:464–470

Rajen G, Kulacki FA (1987) Natural convection in a porous layer locally heated from below—a regional laboratory model for a nuclear waste repository. In: McAssey EV et al (eds) Heat transfer problems in nuclear waste management, American Society of Mechanical Engineers, New York, HTD-67:19–26

Rhee SJ, Dhir VK, Catton I (1978) Natural convection heat transfer in beds of inductively heated particles. ASME J. Heat Trans 100:78–85

Steven S (2006) Experimental study of free convection in fluid-porous layer heated from below. University of Minnesota, Minneapolis, Master's thesis

Singh AK, Thrope GR (1995) Natural convection in a confined fluid overlying a porous layer—a comparison study of different models. Indian J Pure Appl Math 26:81–95

Sommerton CW, Catton I (1982) Thermal instability of superposed porous and fluid layers. ASME J Heat Trans 104:160–165

Soong CY, Tzeng PY, Chiang D, Sheu TS (1996) Numerical study on mode transition of natural convection in differentially heated inclined enclosures. Int J Heat Mass Trans 31:2869–2882

Spalding DB (1972) A novel finite-difference formulation for differential expression involving both first and second order derivatives. Int J Num Meth Eng 4:551–555

Straughan B (2002) Effect of property variation and modeling on convection in a fluid overlying porous layer. In J. Num Anal Meth Geomech 2:75–97

Sun WJ (1973) Convective instability in superposed porous and free layers. University of Minnesota, Minneapolis, Doctoral thesis

Vasseur P, Wang CH, Sen M (1989) The brinkman model for natural convection in a shallow porous cavity with uniform heat flux. Num Heat Trans A: Appl 15:221–242

Worster MG (1992) Instabilities of the liquid and mushy regions during solidification of alloys. J Fluid Mech 237:649–669

Yagi S, Kunii D (1957) Studies on effective thermal conductivities in packed beds. AIChE J 3:373–381

Zhao P, Chen CF (2001) Stability analysis of double-diffusive convection in superposed fluid and porous layers using a one-equation model. Int J Heat Heat Mass Trans 44:4625–4683